S. HRG. 108–458

NUCLEAR POWER GENERATION IN THE UNITED STATES

HEARING

BEFORE THE

SUBCOMMITTEE ON ENERGY

OF THE

COMMITTEE ON ENERGY AND NATURAL RESOURCES UNITED STATES SENATE

ONE HUNDRED EIGHTH CONGRESS

SECOND SESSION

TO RECEIVE TESTIMONY REGARDING NEW NUCLEAR POWER GENERATION IN THE UNITED STATES

———

MARCH 4, 2004

Printed for the use of the
Committee on Energy and Natural Resources

———

U.S. GOVERNMENT PRINTING OFFICE

93–750 PDF   WASHINGTON : 2004

CONTENTS

STATEMENTS

Page

Alexander, Hon. Lamar, U.S. Senator from Tennessee 1
Asselstine, James K., Managing Director, Lehman Brothers, Inc. 38
Bernhard, J.M., Jr., Chairman and CEO, The Shaw Group 31
Craig, Hon. Larry E., U.S. Senator from Idaho 20
Domenici, Hon. Pete V., U.S. Senator from New Mexico 3
Fertel, Marvin S., Senior Vice President and Chief Nuclear Officer, Nuclear
    Energy Institute 21
Landrieu, Hon. Mary L., U.S. Senator from Louisiana 18
McCullough, Glenn L., Jr., Chairman, Board of Directors, Tennessee Valley
    Authority, accompanied by Ike Zeringue, President and Chief Operating
    Officer 7
Travers, Dr. William D., Executive Director for Operations, U.S. Nuclear
    Regulatory Commission 11

APPENDIX

Responses to additional questions 53

NUCLEAR POWER GENERATION IN THE UNITED STATES

THURSDAY, MARCH 4, 2004

U.S. SENATE,
SUBCOMMITTEE ON ENERGY,
COMMITTEE ON ENERGY AND NATURAL RESOURCES,
Washington, DC.

The subcommittee met, pursuant to notice, at 2:30 p.m., in room SD–366, Dirksen Senate Office Building, Hon. Lamar Alexander presiding.

OPENING STATEMENT OF HON. LAMAR ALEXANDER, U.S. SENATOR FROM TENNESSEE

Senator ALEXANDER. The hearing will come to order.

I want to recognize the presence of and thank Chairman Pete Domenici for being here and allowing me to hold this hearing. Senator Domenici has been a leader, the leader really, in the Senate in terms of trying to focus on today's subject, which is the future of generating electricity by nuclear power.

I want to thank Chairman Jim Inhofe for allowing the Nuclear Regulatory Commission to be a part of this hearing. That is a Senate jurisdictional matter and we are grateful for that.

Senator Domenici, would you like to go ahead with your statement or would you like——

The CHAIRMAN. If you have one, why do you not go ahead and then I will follow.

Senator ALEXANDER. I have one I would like to make. Then we will go to yours. I know other Senators are coming and I have plenty of questions if they do not.

This is a hearing on the future of nuclear power generation in the United States. The Tennessee Valley Authority is rebuilding unit 1 of its nuclear power plant at Browns Ferry, Alabama, which has been closed since 1985. This will be the first nuclear capacity in the United States since 1996 when TVA started operations at Watts Bar in Tennessee. There has been no nuclear power plant built from scratch in the United States since 1974.

On the face of it, the failure of the United States to begin a new nuclear plant for 30 years is perplexing. After all, we invented the technology. Since the 1950's we have operated nuclear powered submarines and carriers without a reactor incident. 72 nuclear powered submarines and 9 nuclear powered aircraft carriers operate today all over the world. On shore in the United States, we operate 103 nuclear power plants, which produce about 20 percent of America's electricity.

(1)

At a time when we are importing nearly 70 percent of our oil and increasingly more of our natural gas, our failure to use more nuclear power makes us more dependent on the Middle East and other foreign sources of natural gas. At a time when our supply of natural gas is diminishing and the price is skyrocketing, new nuclear power plants, once built, could provide low-cost electricity that would help keep production costs down and keep jobs from moving overseas. We hear a lot of talk in the U.S. Senate about jobs moving overseas. A good solution would be to talk more about how we can produce more energy to keep production costs down so jobs can stay in the United States. Once built, a nuclear power plant produces electricity at a cost of 1.71 cents per kilowatt hour compared with 1.85 from coal-fired plants and 4.06 from natural gas plants.

Nuclear power plants are efficient and reliable. Coal plants operate 69 to 70 percent of the time; nuclear plants typically, 90 percent of the time.

Finally, at a time when many parts of the United States are struggling to meet clean air standards, nuclear power plants could produce electricity without producing the millions of tons of sulfur dioxide, nitrous oxides, and carbon that coal-fired plants produce. I am particularly sensitive to that because Tennessee has a serious clean air problem, and I notice that in some of the testimonies today, that point is mentioned. It is a subject I would like to discuss more. I think the advantages of nuclear power for cleaning our air are not as well understood as they need to be.

During the 30 years that the United States has not built new nuclear power plants, other countries have. France produces 78 percent of its electricity from nuclear power. Japan, once devastated by nuclear weapons, generates one-third of its electricity from nuclear power. It has three new nuclear power plants under construction. There are eight new nuclear power plants under construction in India, four in China, three in Russia.

America's failure to use this clean, efficient, and inexpensive source of power has been caused by safety failures, ineffective regulation, and poor management decisions. During the 1970's, the industry was mired in cost overruns and schedule delays, brought in part by a changing regulatory environment and management decisions that overestimated needs for capacity.

Nowhere was this more pronounced than at the Tennessee Valley Authority. During the 1970's, TVA had a plan to build 17 nuclear power plants. Four of these were canceled in 1982. Four more were canceled in 1984. The total investment in these eight was $4.6 billion. TVA spent 12 years then building the Sequoyah Plant and 23 years completing Watts Bar, both of which operate today in Tennessee.

But when you add in the costs of interest on these billions spent on unused nuclear capacity, you get an amount that probably equals half—some say more—of TVA's current $24 billion, $25 billion, or $26 billion debt, a debt that raises rates for residents, discourages businesses, and limits TVA's ability to meet its clean air responsibility.

A fire at Browns Ferry in 1975, the Three Mile Island incident in 1979, at which no one was hurt, added to the blows to confidence in the safe, efficient operation of nuclear power plants.

What we hope to find out today is this: whether attitudes and conditions have changed. There does seem to be a regulatory environment that is more favorable to the safe and efficient operation of nuclear power plants. The Nuclear Regulatory Commission, from whom we will hear today, has approved 100 power uprates at existing nuclear plants. They have granted 23 operating license extensions and many more applications are pending. There are new ways to reduce construction costs of the plants. We will hear about that today.

Japan has now constructed the world's first advanced boiling water reactor. Japan did this in a total of 37 months from breaking ground to the loading of the fuel in the reactor core. A U.S. company designed the new reactor. A U.S. regulatory commission certified it, and the Japanese built it and use it.

And the NRC, our regulatory commission, has now established a clear process for early site permitting for new plants. Still, high and uncertain construction costs and lingering safety concerns, along with the Enron debacle, continue to make investors wary of nuclear power plants. The result has been that 90 percent of America's new generating capacity now comes from new plants which burn natural gas, despite higher prices.

I have a chart today that shows that, and I want to underscore that because I hope, Chairman Domenici, we will get back to it. Here we are sitting on 500 years or more of coal, having invented nuclear power, with natural gas prices skyrocketing and over the last 10 years, that is about all we are using to create new plants for generating electricity capacity.

That is why the Nation is so closely watching the progress of the Tennessee Valley Authority at Browns Ferry. If the Browns Ferry project is successfully completed on time and on budget, it could have an impact on the willingness of other utilities and other financial institutions to invest in nuclear power.

In addition, today we hope to learn more about advances in reactor technology and engineering and construction practices. We want to discuss the regulatory framework under which the nuclear industry operates and the financial impediments to developing new nuclear generation in the United States.

I want to thank the witnesses for coming. I will introduce each of you in just a moment, but first I want to turn to Chairman Pete Domenici for his opening comments.

STATEMENT OF HON. PETE V. DOMENICI, U.S. SENATOR FROM NEW MEXICO

The CHAIRMAN. Well, thank you very much, Senator Alexander. Welcome to all of you. It is truly a pleasure to be here with you.

It is not very often that we have a situation in the U.S. Senate where there is only one meeting going on at the same time, and nobody bothers to ask you, when they set a new set of hearings, whether you are busy somewhere else. So that is too difficult even with all the modern technology we have got.

So we have the budget being marked up on the sixth floor of this building. I took this committee over after 24 years of that, maybe 18 or 20 years as chairman of that other committee. Since I am still on it, I have to go up there and vote.

But my real concern now and real love is the U.S. energy situation and why we have done so many dumb things. What can we do to change these around? Well, we will get a little more enlightened.

I have a prepared statement. I do not think I am going to give it because it parallels in many ways your statement. I would like it to be made part of the record and just talk with you all a bit.

Anybody in the United States that is in the energy business that can take a positive step in the production of nuclear energy is taking a giant step toward America's independence in terms of energy that we use for our future.

If you look around, it is energy and more specifically electricity that moves America. And what we have done after inventing it, after creating the greatest engineers, the greatest models, in terms of nuclear power, we all of a sudden got scared. Frankly, if historians worked hard enough at going into the background, we would find that we brought it on ourselves. We produced regulatory schemes that were destined to failure. We produced litigation potential in every filing that were bonanzas to attorneys and where bushel baskets of regulations dropped on the door of anybody that was involved in the system, only to find, when they got through it all, that they were still in court. Some of them had invested their money for 13 years.

Then in this great country of ours, a couple of them got through all of that, and somebody would find that yet there was another problem. So in the great State of New York, we did all that and found that we might have to close one down because we did not have an emergency plan, as if billions just grew on trees. It turned out we are pretty powerful, so we did burn a lot of it up and throw it away.

Now, we have an energy bill that is still languishing here. I have got to tell you if there is anything that this Energy Committee did, it was say to America let us get passed here a proposal that over the next decade will bring natural gas on as much as it can, that will clean up coal and use it, that will give some production tax credit to three or four nuclear power plants, which will then go before our modern-day Nuclear Commission and hopefully set the trend with the building of 3,000 or 4,000 megawatts of nuclear power plants. It has been this Senator's feeling that if just that much happened, it would change things.

Now, Mr. Chairman, or acting chairman and subcommittee chairman, I might say that it is kind of ludicrous. What is really holding it all up was the regulatory scheme. I think to the chagrin of the antis, it has kind of fixed itself up. I take a great deal of pleasure in having a little bit to do with that. It is much better today than it was about 12 years ago, and if one wants to look at history, they will find somewhere or another in that period of time, some chairman got involved. His name was Domenici and it was a committee that paid for your work. All of a sudden we found that you were not going to get paid because you were not doing your work. Boy, did you find out what you were doing wrong fast. Otherwise, you would have had half of the force you had before. What a change since then.

Now, what remains, believe it or not—and, Mr. Chairman, the great new Senator from Tennessee—what holds America today now is we do not know what to do with the waste disposal from the nuclear power plants. It is nigh on an absurdity to be holding up nuclear power because we cannot make up our mind what to do with it.

That is why I am very pleased that you talked about the seas of the world, the oceans. I have that in a couple of speeches. You might have been sitting in a chair once when I gave it. But what I did was I had somebody say on that day where are all the nuclear engines in the waters of the world, and then we found how many used reactors were floating around in oceans and bays and harbors. And we found that no city, except one in New Zealand, precluded those spent fuel rods on board those American ships from being docked right there. Where? Right there in the middle of thousands of ships and millions of people.

And we are running around here saying we cannot even move nuclear power 100 miles down the road because something might happen. They are moving it thousands and then ending up in a port in Italy with an aircraft carrier that has not one nuclear power plant, but two.

So if you can tell that I am worked up about it, I am because I believe we ought to see a decade when four or five new ones are built. And those red lines—they got to turn around pretty soon because there is no more gas. Every plant around has contracted for gas and we are producing all we can. But you know we are going to be using LNG. Is that not interesting? Just think of that. 15 years ago, 20, we said, why are we going overseas for all our crude oil? We never should do that. Now we are using so much natural gas, we are going to run out of it. We are using it for power plants, and they are saying, well, it might work out. We might use Algerian natural gas. It is still not ours. Right? It is still a way-off country somewhere over there.

We will have to be worried when one of their ships blows up. We never thought about it. The other day one of them blew up in a harbor and killed some people, which we are sorry for. It reduced the supply of natural gas through LNG sufficiently that the American market responded. That is interesting. Just think of the state of affairs.

I am so pleased you all as experts are here because we are going to learn a lot from you, but my biggest hope is that we do not only learn, but that you proceed to get some things done. I hope you do. We are going to be here encouraging you.

Thank you very much for the hearing. I hope everything goes well. In the meantime, I will find a little time to go upstairs on the budget, and if you need me because you have to leave——

Senator ALEXANDER. I will be here.

The CHAIRMAN [continuing]. Just call me and I will come from the other hearing.

Thank you all very much.

[The prepared statement of Senator Domenici follows:]

PREPARED STATEMENT OF HON. PETE V. DOMENICI, U.S. SENATOR
FROM NEW MEXICO

I want to first thank Senator Alexander for convening this hearing of his Subcommittee to discuss New Nuclear Power Generation in the United States. This is a critical topic—one in which I've been very involved.

This morning the full Committee heard the EIA Fuel Forecasts. Their forecast calls for 23 percent imported natural gas by 2025, and for 70 percent imported petroleum by the same date.

Compounding these issues is the chart that Senator Alexander showed. I've used this chart several times on the Senate Floor to make the point that we are becoming increasingly reliant, in my view far too reliant, on natural gas for production of electricity. At the same time, natural gas prices are so high that we are losing industries which depend on reasonably priced natural gas as a chemical feedstock.

Nuclear energy provides 20 percent of our electricity today, and does it without emission of pollutants. Our nuclear plants continue to set new records for reliable, low cost performance and their safety record is superb. But it's been so long since we've built a new nuclear plant in the U.S. that we are close to losing the national infrastructure and capability to expand production in the nuclear sector.

Too many times in our history, we've seen our limited diversity of energy sources and our dependence on foreign supplies damage our economy. That's why I've argued that nuclear power must remain a credible option for our future energy portfolio. And that's why I'm pleased that Senator Alexander has called this hearing.

My reasoning on nuclear power is shared in Europe. Just in the last few days, an influential European group, the European Economic and Social Committee, issued an important new Opinion on "nuclear power and electricity generation." The vote on this Opinion was 61 percent in favor with 10 percent abstaining.

That new European Opinion spells out strong support for the future of nuclear power in Europe. Just to quote some of their conclusions, they note that

- nuclear power must be one of the elements of a diversified, balanced, economic and sustainable energy policy with the EU,
- nuclear power is essential if the EU is to successfully apply the concept of sustainable development in policy making, and
- abandoning nuclear power would exacerbate the problems associated with climate change.

I'm pleased to see these recent conclusions from Europe. They are equally applicable here. That's why the comprehensive energy bill that I've been developing for the last year encourages new plant construction in the U.S.

H.R. 6 proposes production tax credits for the first few nuclear plants to encourage utilities to undertake the financial and regulatory risks that will be associated with construction in the U.S. after so long a hiatus. My hope is that construction of those plants will demonstrate to the utilities and the financial community that these plants can be built here with the same success that has been recently demonstrated overseas. In addition, by building a few plants, the construction companies should get past the higher costs associated with "first-of-a-kind" plants and be able to reliably predict and demonstrate lower prices for subsequent plants.

But given the demands that I noted earlier for natural gas, there's another benefit to the nuclear production tax credit that should be emphasized. The EIA projects that even if the only effect of the tax credit is to build the first 6 plants, just those 6 plants will lead to a 3 percent reduction in usage of natural gas. Of course, my goal is that building those first 6 plants leads to far more construction without any government help.

But even if gas prices have only risen to $4 per thousand cubic feet by 2020, which most people would say is far too low an estimate, that 3 percent gas savings translates to a savings to the American public of $3.6 billion per year. Over the 40 to 60 year life of a new plant, that's a lot of billions of savings. And since we'll be importing a large fraction of that natural gas by 2020, that savings translates directly into an improvement in our balance of payments.

I'm pleased that this recent EIA analysis supports the importance of the nuclear production tax credits, and I hope we have that comprehensive energy bill in law before too much longer.

Through this hearing, I hope we can understand the TVA plans at Browns Ferry, which are certainly an important step for the nation along the road to future new nuclear capacity. And I also hope we will better understand the opportunities and challenges for construction of new nuclear plants for America.

Senator ALEXANDER. Thank you, Mr. Chairman, for your comments and your leadership.

Here is how I would like to suggest we proceed. All the Senators have your testimony and have read it or looked it over. I would like to ask you to summarize your statements in about 5 minutes, if you could, which will leave more time for questions. Then I will ask the Senators, when they come, to summarize their remarks in about 5 minutes.

Let me mention all of the witnesses now, and then we will start with Mr. McCullough and just go down the line. Glenn McCullough, Jr. is Chairman of the Board of Directors of the Tennessee Valley Authority, and Ike Zeringue, who is President and Chief Operating Officer of the TVA, is also with him today. We welcome you. Marvin Fertel, senior vice president and chief nuclear officer of the Nuclear Energy Institute. James Asselstine, managing director of Lehman Brothers. Dr. William D. Travers, Executive Director of Operations, Nuclear Regulatory Commission. And James Bernhard, chairman and CEO of The Shaw Group.

Now, what we will do is I will ask Mr. McCullough to start, and may we go just right down the line. I hope you will permit me a little bit of an editorial comment since the Tennessee Valley Authority is headquartered in Knoxville and serves most Tennesseans. We are very glad to have Mr. McCullough here today. He knows very well that as one Senator, I fully support TVA's decision to move ahead with Browns Ferry. Everything I can tell about it suggests to me that they are doing a superb job working with contractors. I hope to find out today what I believe is true, that it is on schedule and on budget. I hope it stays that way and that it serves as a good example to the rest of the country and encourages others to emulate that.

One reason I am so interested in this, beyond the fact that it is an efficient way of producing electricity, is because on April 15, about 75 percent of Tennesseans will be living in counties that are in violation of the clean air standards in our State, and the nuclear power plants, as will be developed in the testimony today, are enormous contributors to clean air. I think we may hear more about that from Mr. Fertel in his testimony, maybe from others of you.

But I congratulate TVA for its gutsy decision to move ahead. The country is watching you, and we look forward to your testimony. Thank you, Mr. McCullough.

STATEMENT OF GLENN L. McCULLOUGH, JR., CHAIRMAN, BOARD OF DIRECTORS, TENNESSEE VALLEY AUTHORITY, ACCOMPANIED BY IKE ZERINGUE, PRESIDENT AND CHIEF OPERATING OFFICER

Mr. McCULLOUGH. Thank you, Mr. Chairman and members of the subcommittee. Thank you for your passion for energy security for this Nation, for a cleaner environment, for a strong economy that produces jobs for the people of America. Also, we appreciate your focus on TVA's nuclear program and the restart of Browns Ferry unit 1.

I am Glenn McCullough, Jr., as you have noted, Mr. Chairman, of the TVA board. With me is Ike Zeringue. Ike is TVA's President and Chief Operating Officer.

Consistent with President Bush's National Energy Policy, my fellow TVA Directors, Skila Harris and Bill Baxter, and I are committed to the continued development of TVA's nuclear program. We are going to focus on maintaining TVA's balanced portfolio of fuel sources, using it as a cost effective source of baseload power and supporting TVA's commitment to cleaner air.

TVA, as you have noted, has a well-balanced portfolio of energy sources that includes nuclear, hydro, coal, and natural gas. This portfolio reflects the Nation's energy mix and it minimizes the price and availability risks that are associated with an overdependence on a single energy source. TVA's three nuclear plants, Browns Ferry, Sequoyah, and Watts Bar, consistently rank among the Nation's most efficient generators.

The cost effectiveness of nuclear power was brought home to the TVA Board during the process of deciding to restart Browns Ferry 1. The 21st century forecast indicated that additional baseload generation would be needed to meet the growing energy demands of 8.5 million people who depend on TVA each day for their electric power. The TVA studied several potential options which included combined cycle natural gas turbines, coal gasification, completing one of the deferred nuclear units at the Bellefonte site and the recovery of Browns Ferry unit 1.

We studied each of these options in terms of fuel price stability, long-term production costs, the environmental impact, potential impact to TVA's long-term ability to reduce debt, capital costs, and estimated capacity factor for meeting the baseload needs. Our study showed clearly that Browns Ferry 1 would produce the needed energy at a very competitive rate while optimizing the value of an existing asset.

Financial analyses also indicated that the operation of all three units at Browns Ferry over an extended license period could reduce TVA's delivered cost of power relative to the market, as well as give TVA more financial flexibility for the future.

Accordingly, the board issued a record of decision in May 2002, authorizing work to begin to return Browns Ferry 1 to service.

It is anticipated that the Browns Ferry 1 recovery project will cost approximately $1.8 billion excluding allowance for funds used during the construction period. TVA is funding the restart of Browns Ferry 1 through existing cash from operations.

The Browns Ferry project is performing to plan. It is now more than 41 percent complete in meeting its 60-month baseline schedule. The project, as you have noted, is on budget.

When unit 1 returns to service, it will provide 1,280 megawatts of cost effective, emission-free generating capacity and will help TVA responsibly meet the growing power demands across the Tennessee Valley.

Finally, in addition to the positive financial benefits, the Browns Ferry 1 restart supports TVA's clean air strategy. The reason is simple. Nuclear power plants do not burn fossil fuel so they do not emit combustion products such as sulfur dioxide and nitrogen oxides into the atmosphere.

Restarting Browns Ferry unit 1 is a beneficial business investment for TVA, for our customers, and for the people of the Tennessee Valley. It will provide clean, affordable, reliable power that will enable the region to continue on a path of economic progress.

Again, thank you for this opportunity for Mr. Zeringue and I to be here. We look forward to answering your questions.

[The prepared statement of Mr. McCullough follows:]

PREPARED STATEMENT OF GLENN L. MCCULLOUGH, JR., CHAIRMAN, BOARD OF DIRECTORS, TENNESSEE VALLEY AUTHORITY

OPENING

On behalf of the Tennessee Valley Authority, I would like to thank you for the opportunity to be here today to discuss TVA's nuclear program and the Browns Ferry Unit 1 restart. My name is Glenn L. McCullough, Jr. I have served on the TVA Board of Directors since November 1999, and I was designated Chairman by President George W. Bush on July 19, 2001. TVA exists to serve the needs of its 158 power distributors and 62 directly served customers and the 8.5 million people of the Tennessee Valley by providing affordable and reliable electric power, environmental stewardship, and leadership in sustainable economic development. A corporation of the federal government, TVA is entirely self-financing and receives no funding from Congress.

TVA is committed to conducting its business in an open and forthright manner that earns the confidence of Congress and the Administration, and in our customers, our investors, and the people of the Tennessee Valley.

HISTORY OF TVA NUCLEAR

TVA's commitment to meeting the region's electricity needs while protecting the environment and supporting a vibrant economy is consistent with President George W. Bush's National Energy Policy. TVA maintains a diverse fuel mix and a strong national transmission system. TVA's strategy of investing in a balanced portfolio of energy sources—nuclear, hydro, coal, natural gas—is similar to the nation's energy mix, and minimizes the price and availability risks associated with over-dependence on a single energy source

TVA made its commitment to nuclear power in the late 1960s and early 1970s, when power sales were growing at a steady rate. In the Tennessee Valley, the number of electricity customers rose to over 2 million in the 1960s and about 30 percent of all the homes were heated with electricity. At that time, TVA was experiencing an annual growth rate of about 8 percent in demand for electricity, and our forecasts through the mid-1970s were showing continued high growth in demand. TVA, and others in the utility industry, predicted that new generating capacity was needed to satisfy its forecast demand. To meet that need, TVA embarked on an ambitious nuclear power plant construction program. Beginning in 1966, TVA announced plans to build 17 nuclear units at seven sites in Tennessee, Alabama, and Mississippi. In 1967, TVA began building the nation's largest nuclear power facility-Browns Ferry in north Alabama.

However, instead of increasing, electricity consumption declined in the mid-1970s following the 1973 energy crisis and again in the late 1970s and 1980s as a result of higher energy costs and slower economic growth. Also, after the Three Mile Island nuclear accident in 1979, the Nuclear Regulatory Commission issued extensive new safety regulations that applied to all plants, whether operating or under construction. This decreasing demand for electricity, coupled with the increased costs of regulation, caused the electric utility industry to rethink the role that nuclear power would play in meeting the nation's demand for electricity. By the early 1980s, TVA and many other utilities had chosen to cancel several nuclear plants that were either planned or under construction.

In 1985, TVA voluntarily shut down all of its operational nuclear units to address regulatory and management issues. TVA implemented a strong performance improvement program, and began returning these units to operation in 1988 with the restart of Sequoyah 1 and 2. Browns Ferry Units 2 and 3 returned to service in 1991 and 1995 respectively, and TVA brought the last licensed U. S. nuclear unit online with Watts Bar 1 in 1996. Based on forecasted baseload power needs at that time, TVA elected not to return Browns Ferry Unit 1 to service with the other units.

TVA NUCLEAR TODAY

TVA's nuclear power program now ranks among industry leaders, in both cost and reliability.

In fiscal year 2003, nuclear power represented about 20 percent of TVA's installed capacity, and produced about 29 percent of TVA's generation.

All three of TVA's nuclear power plants—Browns Ferry, Sequoyah and Watts Bar—rank among the most efficient generators in the country for 2002 and over the past three years, according to Platts Nucleonics Week. TVA is the only utility listed with three plants among the top 15 most efficient generators for 2002 and for the three-year period of 2000-2002.

Sequoyah earned the title of the most efficient generator in the country by producing power at 11.48 mills per kilowatt-hour from 2000-2002. Browns Ferry comes in second at 12.06 mills/kwh, and Watts Bar ranks 12th at 14.39 mill/kwh. In order to achieve these low rates, TVA continues to focus on cost containment through continuous process improvement, standardization, and resource sharing. These efforts resulted in savings of 5.2 million dollars in FY 2003 alone.

Last year TVA received the Nuclear Energy Institute's Top Industry Practice "Best of the Best" Award for strategic planning programs and processes at TVA nuclear plants.

BROWNS FERRY 1 RESTART

Browns Ferry is a three-unit nuclear power facility located just west of Huntsville, Alabama. The plant is owned and operated by TVA to produce electricity for our service area.

As TVA entered the 21st century, forecasts indicated that additional baseload generation would be needed to meet the growing energy demands of our 8.5 million customers. To meet this need, TVA studied several potential options, including combined-cycle gas turbines, coal gasification, startup of one of the deferred Bellefonte nuclear units, and the recovery of Browns Ferry Unit 1.

Each option was studied in terms of fuel price stability; long-term production costs; environmental impact; potential impact to TVA's long-term ability to reduce debt; capital cost; and estimated capacity factor for meeting baseload needs.

After completing these studies, it became clear that Browns Ferry Unit 1 would be able to produce the needed energy at very competitive rates as compared to the other available options, while optimizing the value of an existing asset.

To ensure that a fully informed decision could be made, the TVA Board requested in September 2001, that a detailed scoping study be performed to determine the cost and schedule for recovering and restarting Browns Ferry Unit 1.

TVA also conducted an 18 month environmental impact study under the National Environmental Policy Act to assess not only the restart of Browns Ferry 1, but also the power uprate and potential license renewal for all three units.

The detailed scoping study was nearly unprecedented in the level of detail reviewed to identify not only the remaining work to be done, but also to identify any risk to cost or schedule. Along with the completed environmental study, it provided the TVA Board with comprehensive information to make future base load generation decisions.

Financial analyses also indicated that the operation of all three units at Browns Ferry, over an extended license period, could reduce TVA's delivered cost of power relative to the market, giving TVA more financial flexibility for the future.

Accordingly, the Board issued a record of decision in May 2002, authorizing work to begin to return Browns Ferry Unit 1 to service.

It is anticipated that the Browns Ferry Unit 1 recovery project will cost approximately $1.8 billion, excluding allowance for funds used during construction. TVA will fund the restart of Browns Ferry Unit 1 through existing cash from operations.

The Browns Ferry Unit 1 restart project continues to perform to plan and is more than 40-percent complete in meeting its 60-month baseline schedule. The project remains on budget, with expenses of 381 million dollars last year and about 365 million dollars planned for FY 2004. Planned spending for fiscal year 2005 is 419 million dollars, with spending declining to 381 million in fiscal year 2006, and 129 million in 2007.

When Unit 1 returns to service, its cost-effective and clean, emission-free generating capacity of 1,280 megawatts will help TVA responsibly meet growing power demands while maintaining a strong reserve margin. Our current resource-planning analysis shows that this nuclear unit will help us meet our growing energy needs at a very competitive cost by reducing our delivered cost of power by about .09 cents per kilowatt-hour in its first year of operation.

CLEAN AIR BENEFITS

In addition to the positive financial benefits that the Browns Ferry Unit 1 restart will provide, nuclear plants also support TVA's clean air strategy. The reason is simple nuclear power plants do not burn fossil fuel so they don't emit combustion products such as sulfur dioxide and nitrogen oxides into the atmosphere.

TVA will continue to participate in industry studies of environmentally sound and cost effective power generation technologies for our customers' future energy needs, because it is the right thing to do. Nuclear power remains a vital part of TVA's and the nation's energy portfolio not only because of its clean air benefits, but also because of its strong operational performance.

CONCLUSION

TVA is committed to providing low-cost, reliable power by operating and maintaining safe and efficient plants, standardizing processes across the organization, and continuously improving all aspects of performance. Restarting Browns Ferry Unit 1 is a wise business investment for TVA and our customers. It will provide clean, affordable, and reliable power, enabling TVA to meet the future power demands of the Tennessee Valley.

Senator ALEXANDER. Thank you, Chairman McCullough.

Dr. Travers.

STATEMENT OF DR. WILLIAM D. TRAVERS, EXECUTIVE DIRECTOR FOR OPERATIONS, U.S. NUCLEAR REGULATORY COMMISSION

Dr. TRAVERS. Thank you, Mr. Chairman. It is a pleasure to be here today as you consider new nuclear power generation in the United States. Because, as you are aware, the role of the Nuclear Regulatory Commission as prescribed by the Congress is safety regulation and not promotional, my discussion will focus on actions the Commission has taken and is taking to ensure the continued safe application of nuclear technology, to strengthen regulatory predictability, and to facilitate public access to our information and participation in our processes.

Let me very briefly describe aspects of our national programs for nuclear plant license renewal and power uprates, and then I will outline some of our activities specific to the Browns Ferry nuclear plants, and lastly I will touch on what we have been doing to prepare for the possibility of new reactor licensing in the United States.

First, license renewal. With the improved economic conditions for operating nuclear power plants, the Commission has seen a sustained, strong interest in license renewal which allows plants to operate up to 20 years beyond the original 40-year operating license. The focus of the Commission's review of the license renewal applications is on maintaining plant safety, with the primary concern directed on the aging effects of important systems, structures, and components.

The review of a renewal application proceeds along two fundamental paths. One is for the review of safety issues and the other is to assess potential environmental impacts. Applicants must demonstrate that they have identified and can manage the effects of aging and can continue to maintain an acceptable level of safety during the period of extended operation. The applicant must also address the impacts on the environment from extended operation.

To date, as you have indicated already, renewed licenses have been issued for 12 sites, totaling 23 units. The reviews of these applications were completed on or ahead of schedule, which is indicative of both the care exercised by the licensees in preparing their application and on the Commission's emphasis on planning and completing those reviews on schedule.

Applications to renew the licenses for an additional 11 sites, totaling 19 units, are currently under review, which include the application to renew the licenses for Browns Ferry units 1, 2, and 3.

If all of the applications currently under review are approved, approximately 40 percent of the plants in the United States will have had their operating licenses renewed for 20 years. Based on industry statements, the Commission expects essentially all sites to apply for license renewal.

I should note that public participation is an important part of our license renewal process. There are several opportunities for members of the public to question how aging will be managed during the period of extended operation. Concerns may be litigated in an adjudicatory hearing if an adversely affected party appropriately requests a hearing.

We believe the Commission has established a license renewal process that can be completed in a reasonable period of time with clear requirements to assure safe operation for up to an additional 20 years of plant life.

Turning to power uprates, the NRC carefully reviews requests to raise the maximum power level at which a plant may be operated. These increases to a maximum license power level are called power uprates. Power uprates range from requests for small increases of just a few percent based on the recapture of power measurement uncertainty to larger requests, up to 15 or 20 percent, that require substantial hardware modifications to the plants. The focus of NRC's power uprate review is safety, of course, and in all instances, the NRC must be satisfied that adequate safety margins are maintained.

To date, as you have indicated already, the NRC has approved 100 power uprates, which have added approximately 4,100 megawatts electric to the Nation's electric generating capacity. This is the equivalent of about four large operating nuclear power plants.

Currently the NRC has five power uprate applications under review and expects to receive an additional 25 applications through the calendar year 2005. Again, this would add approximately 1,700 megawatts electric to the Nation's electric energy generating capacity.

I would like now to discuss NRC's oversight of the recovery of the Browns Ferry unit 1 plant. The Tennessee Valley Authority, or TVA, is the NRC-regulated licensee for Browns Ferry in Decatur, Alabama. The Browns Ferry site has three essentially identical boiling water reactors designed by General Electric, and all three Browns Ferry units were voluntarily shut down by TVA in March 1985 to address performance and management issues. Following the shutdowns, TVA specified a number of corrective actions which would be completed prior to restart and confirmed TVA's commitment not to restart without NRC's concurrence.

All three units have retained their operating license during this period, and the restart of units 2 and 3 were both completed a number of years ago. Unit 2 was restarted in 1991 and unit 3 was restarted in November of 1995.

Prior to restart of these units, the NRC completed a significant number of inspections and monitoring to assure that TVA had adequately corrected the issues that caused the shutdown. Since restart, NRC's safety inspections have confirmed that TVA has operated units 2 and 3 safely.

On May 16, 2002, as indicated already, the board of directors approved returning unit 1 to service and authorized TVA to request renewal of the existing 40-year operating license. In anticipation of that decision, the NRC initiated early efforts to establish a methodology and process for the oversight of that effort. The Commission has been especially aware of the need to oversee the unique project with openness and public participation as a priority. We plan to provide numerous opportunities for the public to better understand our process, status of our activities, and most importantly the nature and depth of the NRC safety oversight activities at Browns Ferry. Today we have already had three meetings at Browns Ferry, our region 2 site in Atlanta, and in headquarters and we have developed additional information and access on our public web site.

Let me turn now to a few comments about potential——

Senator ALEXANDER. We would like to try to keep these summaries to about 5 minutes, Dr. Travers.

Dr. TRAVERS. If you will allow me. I apologize for not timing this better.

Future nuclear power reactor licensing positioning by the NRC. While improved performance of operating nuclear power plants has resulted in significant increases in electrical output, as we have already seen, it is expected that significant increases in demands for electricity will need to be addressed by new construction.

As a result, industry interest in new construction in the United States has recently emerged, and for our part, NRC is ready to accept applications for new nuclear power plants. We have been conducting a number of activities. Mr. Chairman, you have already mentioned the regulations that are now in place to govern those activities, and I will not go over those.

I will mention just one quick thing, and that is we are currently reviewing early site permits at three sites in Virginia, Illinois, and Mississippi. Our review of these early site permits, if approved, would be the first time this portion of our licensing process will be tested.

In conclusion—and I appreciate the extra time—the Commission has long been and will continue to be dedicated to its mission to ensure adequate protection of public health and safety, the common defense and security, and the environment. We continue to strive for increases in predictability, efficiency, and public openness in our process.

Thank you very much for this opportunity.

[The prepared statement of Dr. Travers follows:]

PREPARED STATEMENT OF DR. WILLIAM D. TRAVERS, EXECUTIVE DIRECTOR FOR OPERATIONS, U.S. NUCLEAR REGULATORY COMMISSION

INTRODUCTION

Good afternoon, Mr. Chairman and members of the Subcommittee. It is a pleasure to appear before you as you consider "New Nuclear Power Generation in the United States." Because, as you are aware, the role of the Nuclear Regulatory Commission as prescribed by the Congress is regulatory and not promotional, my discussion will focus first on actions the Commission has taken and is taking to ensure the continued safe application of nuclear technology; to strengthen regulatory predictability; and to facilitate public access to our information and participation in our process. We believe that the achievement of all three of those goals will enable others to determine how to use the nuclear energy option.

LICENSE RENEWAL

The focus of the Commission's review of license renewal applications is on maintaining plant safety, with the primary concern directed at the effects of aging on important systems, structures, and components. The review of a renewal application proceeds along two paths—one for the review of safety issues and the other to assess potential environmental impacts. Applicants must demonstrate that they have identified and can manage the effects of aging and can continue to maintain an acceptable level of safety during the period of extended operation. The applicant must also address the impacts on the environment from extended operation.

With the improved economic conditions for operating nuclear power plants, the Commission has seen sustained strong interest in license renewal which allows plants to operate up to 20 years beyond their original 40-year operating license. The original 40-year term was established in the Atomic Energy Act and was based on financial and antitrust considerations, not technical limitations.

The decision to seek license renewal is voluntary and rests entirely with nuclear power plant owners. The decision is typically based on the plant's economic viability and whether it can continue to meet the Commission's requirements. To date, renewed licenses have been issued for 12 sites, totaling 23 units. The reviews of these applications were completed on or ahead of schedule, which is indicative of the care exercised by licensees in preparing their applications and the Commission's emphasis on planning and completing the reviews on schedule. Applications to renew the licenses for an additional 11 sites (totaling 19 units) are currently under review, which includes the application to renew the licenses for Browns Ferry Units 1, 2, and 3. If all of the applications currently under review are approved, approximately 40 percent of the plants in the U.S. will have had their operating licenses renewed. Based on industry statements, the Commission expects essentially all sites to apply for license renewal.

The Commission has established a license renewal process that can be completed in a reasonable period of time with clear requirements to assure safe plant operation for up to an additional 20 years of plant life. To help achieve consistency in the preparation and review of renewal applications, the Commission has issued guidance documents that assist plant owners in preparing license renewal applications and that guide the NRC's review of the applications. Lessons learned from ongoing reviews are documented as they are identified and made publicly available for use by future applicants. These guidance documents provide the framework for an effective, efficient, and technically sound review of renewal applications and help maintain the stability and predictability of the license renewal process.

Public participation is an important part of the license renewal process. There are several opportunities for members of the public to question how aging will be managed during the period of extended operation. Concerns may be litigated in an adjudicatory hearing if an adversely affected party appropriately requests a hearing. A license renewal web site is also available that contains key documents associated with license renewal applications as well as information on the license renewal process, regulations, and guidance documents.

Although the license renewal program has been highly successful, the Commission continues to seek further improvements in the process. Using lessons learned from past reviews, the Commission is pursuing revisions to the renewal process that should provide additional efficiencies. These efficiencies will help the Commission better accommodate the increasing number of renewal applications being submitted.

The Commission recognizes the importance of license renewal to the owners of nuclear power plants and to the future energy needs of the country. The Commission remains committed to providing continued high-priority attention to this effort, while assuring plant safety and maintaining public health and safety.

POWER UPRATE

The NRC carefully reviews requests to raise the maximum power level at which a plant may be operated, which are called power uprates. Improvements of instrument accuracy and plant hardware modifications have allowed licensees to submit power uprate applications for NRC review and approval. The focus of NRC review of these applications has been and will continue to be on safety. We continue to closely monitor operating experience to identify issues that may affect power uprate implementation.

Power uprates range from requests for small increases based on the recapture of power measurement uncertainty, to large requests in the 15 to 20 percent range that require substantial hardware modifications at the plants. In all instances, the NRC must be satisfied that safety margins are maintained. To date, the NRC has approved 100 power uprates which have added approximately 4140 megawatts electric to the nation's electric generating capacity—the equivalent of about four large nuclear power plants.

Currently, the NRC has five power uprate applications under review and expects to receive an additional 25 applications through calendar year 2005. This would add approximately 1760 megawatts electric to the nation's electric generating capacity.

In recognition of the increased interest in power uprates by licensees, NRC recently issued a Review Standard for Extended Power Uprates (i.e., uprates that increase the current power by seven percent or more). This document, which is available publicly, provides a "road map" that establishes standardized review guidance and acceptance criteria for both the NRC and licensees. The Review Standard enhances the NRC's focus on safety and improves consistency, predictability and efficiency of these reviews. The Review Standard will foster improved communications with our stakeholders and licensees.

The NRC is monitoring operating experience at plants that have implemented power uprates. Cases of steam dryer cracking and flow-induced vibration damage affecting components and supports for the main steam and feedwater lines have been observed to occur at some of these plants. The NRC conducted inspections to identify the causes of several of these issues and evaluated many of the repairs performed by the licensees. Currently, we have determined that these issues do not pose an immediate safety concern. The Commission continues to monitor the industry's generic response to these issues and will consider additional regulatory action, as appropriate.

In summary, the focus of NRC review of power uprate applications continues to be on ensuring public health and safety.

BROWNS FERRY UNIT 1 RESTART

I would now like to discuss the NRC's oversight of the recovery of Browns Ferry Unit 1. The Tennessee Valley Authority, or TVA, is the NRC-regulated licensee for the Browns Ferry Nuclear Power Plant located near Decatur, Alabama. The Browns Ferry site has three essentially identical boiling water reactors designed by General Electric. All three Browns Ferry units were voluntarily shut down by the Tennessee Valley Authority in March of 1985 to address performance and management issues. Following the shutdowns, TVA specified corrective actions which would be completed prior to restart and confirmed TVA's commitment not to restart any unit without NRC's concurrence. All three units retained their operating licenses during their respective long-term shutdown.

The restart efforts for Units 2 and 3 were both approximately five years in duration. Unit 2 was restarted in May 1991, and Unit 3 in November 1995, following Commission briefings and NRC Staff approval of restart. Prior to the restart of these units, the NRC completed significant inspections and monitoring to assure that TVA had adequately corrected the issues that caused the shutdown of all three Browns Ferry Units. TVA has subsequently operated the Unit 2 and 3 reactors in a safe and effective manner.

On May 16, 2002, the TVA Board of Directors approved the return of Browns Ferry Unit 1 to service and authorized TVA Nuclear to request renewal of the existing 40-year operating licenses for all three units. In December 2002, TVA submitted its proposed regulatory framework for the Unit 1 restart. Following a public meeting and after TVA modified several areas, the NRC accepted TVA's proposed framework in August 2003. This presents a unique issue of performing a license extension review for a reactor unit that has not been operated for an extensive period of time. However, because license renewal focuses on programs to manage the effects of aging on long-lived components, NRC will be able to provide an effective review of this application. The premise of the application for Unit 1 is that its current license basis is essentially the same as that for Units 2 and 3. In the application, TVA identified differences between Unit 1 and Units 2 and 3 and stated that those differences will be eliminated by Unit 1 restart activities. Through the review of the renewal application, the NRC will identify those contingencies that would be applicable to Unit 1 renewal, such as items that would need to be completed by TVA and included in NRC restart verification activities.

TVA has applied many lessons learned from the restart of the other two units in its recovery plan for Unit 1. One TVA objective for Unit 1 restart is to have all three units "operationally identical" at the completion of the project and to use as many of their current plant processes and procedures as possible. It is important to note that Unit 1 has been maintained by TVA in a "de-fueled lay-up" condition since 1985. Since 1985, the NRC has conducted periodic lay-up inspections to confirm the conditions of key plant components.

In anticipation of the TVA Board's decision to restart Unit 1, the NRC initiated efforts to establish a methodology and plan for oversight of this third Browns Ferry unit recovery and to establish the needed resources. After an extensive review of NRC lessons learned from TVA's recovery of the previous two units and a critical evaluation of differences in TVA's recovery plans for Unit 1, this detailed methodology was formally defined in an NRC Inspection Manual Chapter (MC 2509) issued in August 2003. NRC oversight inspections of the Unit 1 recovery are currently being implemented at the early stages of the recovery process and will be completed for all necessary activities including selected renovation activities, restart testing, and return to operational status.

Based on the TVA plan for restart of Browns Ferry Unit 1 and their use of existing processes which we have previously confirmed as acceptable, the Commission has committed adequate resources throughout the project to support the planned inspection activities. Two additional resident inspectors have already been stationed at the site to provide first-hand monitoring of the licensee's recovery activities. Other staff members have been assigned oversight and specialist inspection roles in our regional and headquarters offices. We have established an experienced team— essentially all of the NRC staff associated with the Unit 1 recovery have been involved in the previous Browns Ferry units recovery efforts or other long-term recoveries. We are using this experience to maximize the effectiveness of our applied inspection resources to ensure the recovery efforts result in a plant that can be operated safely.

The Commission has been especially aware of the need to oversee this unique project with openness to the public as a priority. To facilitate this, a communications plan has been developed which provides for periodic public meetings conducted at a variety of locations. We plan to provide numerous opportunities for the public to better understand the recovery process, status of activities, and most importantly, the nature and depth of the NRC's safety oversight activities at Browns Ferry Unit 1. To date, we have held three such meetings, one in Washington at NRC Headquarters, one in Atlanta at the NRC Regional Office, and one at the Browns Ferry site. In addition, we have developed an easy means for public access to Browns Ferry Unit 1 restart information on our public Web site. The Web site contains information that describes the recovery effort, allows access to our completed inspections, and provides the summaries of the public meetings previously mentioned.

In summary, the Browns Ferry Unit 1 Restart is progressing as planned by TVA, with dedicated NRC inspection, oversight and licensing resources from NRC headquarters and the NRC Region II Office in Atlanta, GA. The NRC has worked effectively with TVA to develop an effective road map for the recovery project to allow for effective and efficient use of both TVA and NRC resources while ensuring our primary safety mission is achieved. The Unit 1 restart effort benefits from the processes established for, and lessons-learned from, the restart of the other two Browns Ferry Units. The Commission has prepared for the increased oversight that a project of this scope warrants, and will continue to work closely with the licensee as the restart effort progresses.

NEW REACTOR LICENSING

While improved performance of operating nuclear power plants has resulted in significant increases in electrical output, it is expected that any significant increased demands for electricity will need to be addressed by construction of new generating capacity. As a result, industry interest in new construction of nuclear power plants in the U.S. has recently emerged. The NRC is ready to accept applications for new nuclear power plants. New nuclear power plants will likely utilize 10 CFR Part 52 which provides a stable and predictable licensing process. This process ensures that all safety and environmental issues, including emergency preparedness and security, are resolved prior to the construction of a new nuclear power plant. The design certification part of the process resolves the safety issues related to the plant design, while the early site permit process resolves safety and environmental issues related to a potential site. The issues resolved in these two parts can then be referenced in an application which would lead to a combined construction permit and operating license with conditions (combined license).

As you know, the Commission has already certified three new reactor designs, pursuant to 10 CFR Part 52, making them readily available for new plant orders. These designs include General Electric's Advanced Boiling Water Reactor and Westinghouse's AP-600 and System 80+ designs.

In addition to the three advanced reactor designs already certified, there are new nuclear power plant technologies which some believe can provide enhanced safety, improved efficiency, lower costs, as well as other benefits. The Commission is currently reviewing the Westinghouse AP1000 design certification application. The staff has met all scheduled milestones for the AP1000 design review and is on track to issue the final design approval recommendation to the Commission this fall, followed by the design certification rule in 2005. The NRC staff is also actively reviewing pre-application issues on two additional designs and has four other designs in various stages of pre-application review.

In September and October of last year, we received three early site permit applications for sites in Virginia, Illinois, and Mississippi where operating reactors already exist. Our review of these early site permits, if approved, would be the first time this portion of the licensing process in 10 CFR Part 52 has been implemented. The staff has established schedules to complete the safety reviews and environmental impact statements in approximately two years. The mandatory adjudicatory hearings associated with the early site permits will be concluded after completion of the NRC staff's technical review. As with the design certification rulemaking, issues resolved in the early site permit proceedings will not be revisited during a combined license proceeding absent new and compelling information.

To prepare for the potential application for a combined license, the Commission is discussing generic issues related to the preparation and review of a combined license application with industry and interested stakeholders. Included in this effort is updating the NRC's construction inspection program to address the combined license process and the expected use of extensive modular construction techniques.

REACTOR OVERSIGHT PROCESS

One vital aspect of our regulatory oversight of commercial nuclear power plants is the direct inspection of equipment and activities at the power plant sites by NRC inspectors. NRC regional inspectors often specialize in areas and perform their specific inspections at the plants throughout the region in which they are assigned. Many of our inspection staff have a number of years experience, both within the NRC and in other parts of the industry, and all are well qualified for their duties. The inspections are conducted in accordance with an agency-wide inspection program. The program defines the frequency and scope of inspection activities and includes detailed inspection procedures, which cover a wide variety of topics, including operations, maintenance, refueling, engineering, radiation protection, emergency planning, and physical security. Our inspection reports are also available to the public, with the exception of those containing sensitive security information.

In addition to our inspection program, the NRC maintains performance indicators to aid in trending the safety performance of the power plants. These performance indicators trend information such as unplanned shutdowns and power changes; the performance of important safety equipment; and control of radiation exposure. There is a baseline level of inspection that all plants receive, and there are increasing levels of additional inspection activities that may be performed if the performance at a given plant indicates it is warranted.

On an annual basis, we assess the performance at each power plant and issue a written summary of our conclusions, followed by a meeting with the operator of the plant and a meeting with members of the local public. The NRC also makes the performance indicators and the results of our inspections available to the public.

The inspection program is coordinated by our staff in the NRC headquarters office in Rockville, Maryland. This coordination ensures consistency in implementation of the inspection program between NRC's regional offices, and aids in the sharing of information within the agency.

SECURITY

Of course with the heightened nuclear security at U.S. commercial reactors, the NRC will ensure that all operating nuclear power plants will be in compliance with all current nuclear security regulatory requirements. In this regard the NRC will continue to coordinate with other federal agencies including the Department of Homeland Security, Homeland Security Council, Federal Bureau of Investigation, and the Intelligence Community to ensure greater awareness of threats and to enhance the communication of threat information from all sources.

SUMMARY

The Commission has long been, and will continue to be, dedicated in its mission to ensure adequate protection of public health and safety, the common defense and security, and the environment in the application of nuclear technology for civilian use. The Commission is mindful of the need to: (1) enhance regulatory predictability and reduce regulatory burden when appropriate and justified, so as not to inappropriately inhibit any renewed interest in nuclear power; (2) maintain open communications with all its stakeholders; and (3) continue to encourage its highly qualified staff to strive for increased efficiency and effectiveness.

I appreciate the opportunity to appear before you today, and I welcome your comments and questions.

Senator ALEXANDER. Thank you, Dr. Travers.

We have been joined by Senator Landrieu, who has a strong interest in nuclear power.

Senator Landrieu.

STATEMENT OF HON. MARY L. LANDRIEU, U.S. SENATOR FROM LOUISIANA

Senator LANDRIEU. Thank you, Mr. Chairman. I welcome our panelists, particularly our friends from The Shaw Group. Thank you, Jim, for being with us today and for participating.

I just have a brief opening statement, but I would like, Mr. Chairman, if I could, just to review some of the comments at this time and not take too long so we can get on with the panel.

I am here because I am a strong supporter of the revitalization of the nuclear industry in our country for any number of reasons. One, because it is very apparent to me and to many on this committee that we need to increase our supply of energy in a diverse and robust supply of energy. Even from a State that produces a lot of oil and gas, not too much coal, we agree as producers that we need to have a greater production of nuclear. There are some real barriers that have made that goal difficult.

So I just want to say for the record any number of things, but specifically that our chemical industry, Mr. Chairman, which is one of the largest industrial users of natural gas and one of the largest employers in my State with Dow Chemical, recently wrote that a diversified and robust energy mix is critically needed. The price of natural gas is so high, it makes our industries in Louisiana very fragile and very difficult, Mr. Chairman, to compete internationally. So we need to seek greater supplies of natural gas from a variety of sources, domestic preferably, but we are also open to importing liquified natural gas. But getting our nuclear industry more robust helps us to increase that supply and in some ways manage the demand.

Number two, nuclear power is the Nation's largest clean air source for electricity and generating today three-fourths of all emission-free electricity. So at a time when our country—and you have been a great leader on the environment—is looking for ways to clean up our air, to meet our new stricter standards for the environment, this is very important.

There are at least six things that we can do immediately, and I will go through them quickly. Restart Browns Ferry unit 1, which will provide 1,254 megawatts of power. Reinstate the 1.7 cent per kilowatt hour production tax credit, which is in this energy bill. That is why it is important for us to pass it. Promote the construction of six plants in a series to reduce the cost of competitive rates from about 1,300 to 1,500 per kilowatt to match gas-fired and coal-fired plants. The Government should share the cost for site banking for a number of the plants. The Government should recognize nuclear as a carbon-free source of energy, and we need to continue to fund the Department's generation IV program for the generation of reactors.

So I am very pleased, Mr. Chairman, to join you today. I am pleased we have Jim Bernhard as one of our witnesses. Jim is chairman and founder of The Shaw Group, one of two Fortune 500 companies in my home State of Louisiana. So we are very proud of the extraordinary growth and tremendous work that this company has done not just in our State, but providing jobs and opportunities for all States of the country and internationally.

Thank you. I look forward to hearing the panelists' discussion.

[The prepared statement of Senator Landrieu follows:]

PREPARED STATEMENT OF HON. MARY L. LANDRIEU, U.S. SENATOR FROM LOUISIANA

Mr. Chairman, today I would like to thank you for convening this hearing on the future of nuclear energy, as a critical energy source in our country today and for the foreseeable future.

The Congress must recognize the important role that nuclear energy plays in our nation's economy, our nation's energy independence and security, and our nation's environmental goals. And, we need to acknowledge that like nearly every other source of energy, nuclear power needs our help to continue playing its important role in our nation's energy policy.

One in every five homes and businesses today is powered by nuclear energy. It is important not only in Louisiana, where two nuclear plants produce nearly 17 percent of my state's electricity, but also in states such as Connecticut, Illinois, New Hampshire, New Jersey, South Carolina and Vermont where nuclear power generates more electricity than any other source. Nationwide, 103 reactors provide 20 percent of our electricity—the largest source of U.S. emission-free power provided around the clock.

Nuclear energy played an important role in the sustained economic growth during the 1990s. By operating more and more efficiently, our nation's nuclear power plants have added the equivalent of twenty-five 1,000-megawatt power plants to our nation's electricity grid. Without that improvement in performance by our nuclear plants, we would have needed at least 25 new power plants; and those plants most likely would not have the clean-air benefits provided by nuclear energy.

While I strongly support the use of natural gas for our energy needs, we cannot rely—as we have in recent years—on only one source of energy to meet our nation's increasing electricity demand. The CEO of Dow Chemical recently wrote that the chemical industry—the nation's largest industrial user of natural gas and one of the largest employers in my state—is particularly vulnerable to high natural gas prices. He advised that the solution is to promote "a diversified and robust energy mix . . . including the full range of traditional and alternative energy sources." If we are going to maintain our world economic leadership, we surely need to follow that advice.

Nuclear energy is also vitally important for our environment and our nation's clean air goals. Nuclear power is the nation's largest clean air source of electricity, generating three-fourths of all emission-free electricity. For future generations of Americans, whose reliance on electricity will increase and who rightfully want a cleaner environment and the health benefits that cleaner air will provide, nuclear energy will be an essential partner.

According to the Department of Energy, the demand for electricity is expected to grow by 50 percent by 2020. In order to continue producing at least one-third of our total electricity generation from emission-free sources, we must build 50,000 megawatts of new nuclear energy production. If we do that, we are just preserving our current levels of emission-free generation, not improving them.

And finally, we need to recognize that nuclear power, by providing a stable and dependable source of electricity, is vital to our nation's energy security and independence. Nuclear power is essentially an American invention. We generate nearly a fourth of the world's total nuclear power and we can do so with domestic energy sources. I agree that hydrogen holds the promise of helping us lessen our dependence on imported oil. Nuclear power is one of the most promising ways that we can produce hydrogen economically and efficiently.

Mr. Chairman, we are at a critical juncture in deciding the future of our energy security and as a result bold steps must be embarked upon to preserve our international competitiveness. We could start by first doing the following:

- Restart Browns Ferry Unit (2 and 3 have already been started) which will provide 1254 Megawatts of power.
- Reinstate the 1.7 cents per kilowatt-hour production tax credit for the constriction of the first 6 "first mover" plants over an 8 year period.
- Promote the construction of these 6 plants in a series so as to reduce cost to a competitive rate of about $1,300 to $1,500 per kilowatt to match gas-fired and coal-fired plants.
- The government should share the cost for site banking for a number of plants, certification of new plant designs by the Nuclear Regulatory Commission, and the combined construction and operation licenses for plants built immediately.
- The government should recognize nuclear as a carbon-free source of energy.
- We need to continue to fund the Department of Energy's Generation IV program for the next generation reactors.

Today, Mr. Chairman, I am pleased that we have Jim Bernhard as one of our witnesses today. Jim is Chairman and CEO, and founder of The Shaw Group, one of two Fortune 500 companies in my home state of Louisiana. Founded in 1987, Shaw has expanded rapidly, through internal growth, and through a series of strategic acquisitions. In fact, it was the acquisition of Stone & Webster three and a half years ago that expanded Shaw's capabilities in the commercial nuclear industry.

Stone & Webster's long history in the nuclear industry dates back to the Manhattan Project, when it built an electromagnetic separation plant that would produce the materials needed for atomic weapons, and also built the City of Oak Ridge, which housed 75,000 workers.

To date, Stone & Webster—now part of The Shaw Group—has designed and/or built 17 nuclear plants and has provided technical services to 95 percent of all U.S. nuclear plants.

With that said, I think it's appropriate that we have The Shaw Group testifying today and I welcome Jim to this hearing. I see that he's brought his lovely wife, Dana, and their two children. Welcome to you also. Thank you, Mr. Chairman.

Senator ALEXANDER. Thank you, Senator Landrieu.

Senator Craig.

STATEMENT OF HON. LARRY E. CRAIG, U.S. SENATOR FROM IDAHO

Senator CRAIG. Mr. Chairman, first of all, let me thank you very much for this hearing. I wish I could stay the whole time. I am not going to be able to, but I will read all of the testimony.

My effort, your effort, all of our efforts to advance the cause of nuclear energy and nuclear technology in this country are probably well known by our panelists.

The work that is going on in my State and across the DOE's national laboratories on advanced nuclear generation is to me very exciting. Much of DOE's nuclear research is focused on what is called generation IV technology. What our witnesses have been talking about today is something called generation III-and-a-half, nuclear technologies that are more advanced than current plants, but could be deployed in the near term than generation IV technology.

I am encouraged that we are examining these near-term opportunities. I think it is important that we keep a focus on the near term while also working on the long term. Many of the current nuclear reactor sites have enough capacity to add additional reactors at that site. We have heard today that Tennessee Valley Authority may have that opportunity.

DOE has also been working with nuclear utilities across the country on what is called early site permitting. The Nuclear Regulatory Commission is also working to modernize and streamline both new plant licenses and also the license extension process for existing reactors. I am encouraged by that and strongly support it.

Let me also add that one of the areas, Mr. Chairman, that I and others have been heavily involved in is the issue of climate change and arguing that it is technology that gets us to where we want to go, not turning off our economies.

I found it fascinating. I happen to be at the climate change conference in Milan in December, and those countries that rushed to judgment by adopting the Kyoto Protocol, feeling that they only had a few percentage points to go to meet their 1999 standards, now are even further out of compliance by another additional amount because what they are finding is quite simple. You cannot advance the economy of your country under current technologies without emitting greenhouse gases. It is the character of current technology, except for one and that is nuclear.

It was interesting to me. Italy, although it had shut one of its nuclear plants when it ratified the Kyoto Protocol several years ago, was within, I believe, 4 or 5 percentage points of being in compliance. It is now out of compliance by 13 percent.

Enough of that. Let us hear from our witnesses.

But there is a reality check out there that I hope America gets right quickly, and that is that if you want your lights to stay on at reasonable costs and if you want your computers to stay on at reasonable costs and if you want to advance all the technologies in our country at reasonable costs, you have got to have quality electricity at reasonable costs. I do not know how you get there without expanding the overall percentage of the blend in our current electrical makeup with nuclear.

Thank you, Mr. Chairman.

Senator ALEXANDER. Thank you, Senator Craig.

Mr. Fertel.

STATEMENT OF MARVIN S. FERTEL, SENIOR VICE PRESIDENT AND CHIEF NUCLEAR OFFICER, NUCLEAR ENERGY INSTITUTE

Mr. FERTEL. On behalf of the Nuclear Energy Institute, Mr. Chairman and Senator Craig, I would like to thank you both for your leadership on national energy issues and particularly for your strong support for nuclear energy.

As I know you are aware, NEI is responsible for developing policy for the U.S. nuclear industry. Our organization's 270 member companies represent a broad spectrum of organizations and interests and includes every U.S. energy company that operates our 103 nuclear power plants in this country.

I will focus my testimony today on four points. First, that nuclear energy is already critical to America's energy supply. Today nuclear power plants provide electricity for 1 in 5 American homes and businesses safely and reliably. As you indicated in your opening statement, Mr. Chairman, nuclear plants are the lowest cost baseload source of electricity, have excellent forward price stability, which helps our consumers, and maybe most importantly, nuclear power plants produce no air pollution or greenhouse gases.

But what we do today is not enough. Our Nation's electricity demands are expected to increase 50 percent by 2025, according to the Energy Information Administration. And we believe that nuclear power plants must play an even greater role in meeting both those needs and our environmental needs as we go forward.

This leads me to my second point. Nuclear energy is an essential part of our diverse fuel mix. Fuel and technology diversity is the core strength of our U.S. electricity supply system. Coal is our largest source of electricity and nuclear is our second largest source. But as your chart shows, natural gas-fired plants account for 95 percent of the new electricity generation built since 1992 which was the last year that a comprehensive energy bill was passed by the Congress.

Now, natural gas is an integral part of our energy mix, but over-reliance on any one fuel source leaves consumers vulnerable to price spikes and supply disruptions, as Alan Greenspan testified last year.

From a national policy perspective, we believe it is essential that nuclear plants, coal plants, and other generation sources are built to satisfy the EIA projections that show a need for 400,000 megawatts of new generation by 2025.

My third point is that the industry is committed to building new nuclear power plants. The demonstration of a new licensing process for building nuclear plants has been underway for 2½ years. As Dr. Travers just indicated, three companies, Dominion Resources, Entergy, and Exelon, filed applications last year with the NRC for early site permits which will allow them to bank sites for new reactors.

Companies are also planning to demonstrate the NRC's new process for obtaining a combined construction and operating license probably later this year. And the industry is investing substantial resources in design and engineering of advanced plants and get them NRC-certified.

And that brings me to my last point and probably the one most pertinent to this committee. Our country needs to stimulate investment in energy infrastructure, including new nuclear power plants. Congress can help create an environment that will stimulate investments in new nuclear power plants through the passage of the comprehensive energy legislation that Senator Domenici spoke so elegantly about earlier. Current legislation contains provisions essential for new plant construction, such as renewal of the Price-Anderson insurance framework, authorization for a cost-share program between industry and government to support the design, engineering, and licensing of new advanced plants. The passage of the energy bill that includes provisions that encourage new plant construction is vital for our country's economic and environmental future.

It is important that Congress send a clear signal to both business decision-makers and the financial community, from which you will hear in a few minutes, that supports investments in new nuclear plants through targeted but limited measures such as the production tax credits that Senator Landrieu mentioned or loan guarantees, measures that the Senate endorsed last year.

Mr. Chairman, we have submitted detailed written testimony. On behalf of NEI, I thank you for allowing me to testify today and look forward to any questions you may have later.

[The prepared statement of Mr. Fertel follows:]

PREPARED STATEMENT OF MARVIN S. FERTEL, SENIOR VICE PRESIDENT AND CHIEF NUCLEAR OFFICER, NUCLEAR ENERGY INSTITUTE

TESTIMONY

Chairman Alexander, Ranking Member Graham and distinguished members of the subcommittee, I am Marvin Fertel, senior vice president and chief nuclear officer at the Nuclear Energy Institute (NEI). NEI appreciates the opportunity to provide this testimony for the record on the need for new nuclear power plants, and the issues that must be addressed before our nation can begin construction of the new nuclear plants needed to meet growing electricity demand in the years ahead.

NEI is responsible for developing policy for the U.S. nuclear industry. Our organization's 270 member companies represent a broad spectrum of interests, including every U.S. energy company that operates a nuclear power plant. NEI's membership also includes nuclear fuel cycle companies, suppliers, engineering and consulting firms, national research laboratories, manufacturers of radiopharmaceuticals, universities, labor unions and law firms.

America's 103 nuclear power plants are the most efficient and reliable in the world. Nuclear energy is the largest source of emission-free electricity in the United States and our nation's second largest source of electricity after coal. Nuclear power plants in 31 states provide electricity for one of every five U.S. homes and businesses. Seven out of 10 Americans believe nuclear energy should play an important role in the country's energy future.[1]

Given these facts and the strategic importance of nuclear energy to our nation's energy security and economic growth, NEI encourages the Congress to adopt policies that foster continued expansion of emission-free nuclear energy as a vital part of our nation's diverse energy mix.

Last year, Congress demonstrated strong support for nuclear energy's role in forward-looking energy policy legislation. That legislation includes most of the major policy initiatives necessary to carry this technology forward into the 21st century as a major contributor to U.S. electricity supply. These include renewal of the Price-Anderson insurance framework; financial stimulus for new nuclear plant construction; an expanded research and development portfolio; support for universities; and updated tax treatment of nuclear decommissioning funds to reflect today's competitive electricity business.

Broadly, the energy sector believes it is imperative to provide substantial stimulus for investment in new transmission infrastructure for both electricity and natural gas, and in the new nuclear and clean coal power plants to meet the 50 percent increase in electricity demand by 2025 forecast by the Energy Information Administration. Investment in key parts of the electric power sector has collapsed over the last 10 years, and we must put in place new policy initiatives to address that challenge.

NEI's testimony for the record will cover the following areas:

1. The business case for new nuclear power plants.
2. Industry initiatives to increase nuclear energy production.
3. The need to stimulate investment in America's critical energy infrastructure, including investment in new nuclear power plants.
4. Industry programs to create the business conditions necessary for the construction of new nuclear plants and the steps to ensure construction of new plants to meet demand for new baseload electric generating capacity.
5. Industry confidence in the competitiveness of new nuclear power plants.

THE BUSINESS CASE FOR NEW NUCLEAR POWER PLANTS

New nuclear plants will be essential in the years ahead to achieve a number of critically important public policy imperatives for our country's energy supply and electricity market.

---

[1] *Perspectives on Public Opinion*, by Ann Stouffer Bisconti, Bisconti Research Inc., November 2003.

First, new nuclear power plants will continue to contribute to the fuel and technology diversity that is the core strength of the U.S. electric supply system. This diversity is at risk because today's business environment and market conditions make investment in large, new capital-intensive technologies difficult, notably the advanced nuclear power plants and advanced coal-fired power plants best suited to supply baseload electricity. More than 90 percent of all new electric generating capacity added over the past five years is fueled with natural gas. Natural gas has many desirable characteristics and should be part of our fuel mix, but "over-reliance on any one fuel source leaves consumers vulnerable to price spikes and supply disruptions."[2]

Second, new nuclear power plants provide future price stability that is not available from electric generating plants fueled with natural gas. Intense volatility in natural gas prices over the last several years is likely to continue, and subjects the U.S. economy to potential damage. Although nuclear plants are capital-intensive to build, the operating costs of nuclear power plants are stable and can dampen volatility of consumer costs in the electricity market.

Third, new nuclear plants will reduce the price and supply volatility of natural gas, thereby relieving cost pressures on other users of natural gas that have no alternative fuel source.

Finally, new nuclear power plants will play a leading role in meeting U.S. clean air goals and the administration's goal of reducing the U.S. economy's greenhouse gas intensity. In addition, under the cap-and-trade systems in place or planned for all major pollutants, incremental production from new emission-free nuclear power plants would reduce the compliance costs that otherwise would be imposed on coal-fired and gas-fired generation.

Nuclear power plants produce electricity that otherwise would be supplied by oil-, gas- or coal-fired generating capacity, and thus avoid the emissions associated with that fossil-fueled capacity.

The value of the emissions avoided by U.S. nuclear power plants is essential in meeting clean air regulations. In 2002, U.S. nuclear power plants avoided the emission of about 3.4 million tons of sodium dioxide (SO_2) and about 1.4 million tons of nitrogen oxide (NO_X). To put these numbers in perspective, the requirements imposed by the 1990 Clean Air Act Amendments reduced SO_2 emissions from the electric power sector between 1990 and 2002 by 5.5 million tons per year and NO_X emissions by 2.3 million tons year.[3] Thus, in a single year, nuclear power plants avoid nearly as much in emissions as was achieved over a 12-year period by other sources.

The NO_X emissions avoided by U.S. nuclear power plants are equivalent to eliminating NO_X emissions from six out of 10 passenger cars in the United States. The carbon emissions avoided by U.S. nuclear power plants are equivalent to eliminating the carbon emissions from nine out of 10 passenger cars in the United States.

EMISSIONS AVOIDED BY NUCLEAR POWER PLANTS

Year	SO_2 (millions short tons)	NO_X (millions short tons)	Carbon (million metric tons of carbon equivalent)
2002	3.38	1.39	1,441.5
Emissions reduced at fossil generating plans 1990-2002 as a result of 1990 Clean Air Act amendments	5.5	2.30	NA

Source: EPA (SO_2 emissions for the electric power sector in 1990 were 15.7 million tons; by 2002, emissions had been reduced to 10.2 million tons, a 5.5-million-ton reduction. NO_X emissions had been reduced to 4.6 million tons, a 2.3-million-ton reduction.

Nuclear energy helped reduce NO_X emissions in northeastern and mid-Atlantic states, according to a report last year by the Environmental Protection Agency and the Ozone Transport Commission (OTC).[4] The 2003 EPA assessment found that energy companies have been shifting electricity production from fossil-fueled power

---

[2] Report of the President's National Energy Policy Development Group, May 2001, page xiii.

[3] *EPA Acid Rain Program: 2001 Progress Report*, U.S. Environmental Protection Agency, November 2002.

[4] *NO_X Budget Program: 1999-2002 Progress Report*, U.S. Environmental Protection Agency, March 2003.

plants to emission-free nuclear power plants to help comply with federal air pollution laws.

In Tennessee, for example, three nuclear reactors avoid the emission of approximately 170,000 tons of SO_2, 60,000 tons of NO_X and 6.6 million metric tons of carbon every year. For perspective, 60,000 tons of NO_X, which is a precursor to ground-level ozone, is the amount released into the air by 3.1 million passenger cars. There are 1.7 million passenger cars registered in Tennessee.

In summary, nuclear energy represents a unique value proposition: a nuclear power plant provides large volumes of electricity—cleanly, reliably, safely and affordably. It provides future price stability and serves as a hedge against the kind of price and supply volatility we see with natural gas. And nuclear plants have valuable environmental attributes: They do not emit controlled air pollutants or carbon dioxide, and thus are not vulnerable to mandatory limits on carbon emissions. Other sources of electricity have some of these attributes. But none of them not coal, natural gas or renewables can deliver all of these benefits. Only nuclear power plants have all of these attributes, and that is why these plants are uniquely valuable.

INDUSTRY INITIATIVES TO INCREASE NUCLEAR ENERGY PRODUCTION

As our country prepares for the construction of new nuclear power plants, the U.S. industry has increased the productivity and efficiency of its existing 103 nuclear power plants.

The industry continues to uprate capacity at U.S. plants—the U.S. Nuclear Regulatory Commission has authorized more than 2,000 megawatts (MW) of power uprates over the last three years, and another 2,000 MW are expected over the next several years. An uprate increases the output of the nuclear reactor and must be approved by the NRC to ensure that the plant can operate safely at the higher production level. Companies will invest in these power uprates as conditions in their local power markets justify.

In addition, energy companies are pursuing renewal of their operating licenses. This option allows today's operating plants to extend their lives for 20 additional years—from 40 to 60 years. Just in the past 12 months, the NRC has approved renewed licenses for 13 reactors, bringing the total number of reactors extending their federal operating licenses to 23.

An additional 33 reactors either have already filed their renewal applications, or indicated formally to NRC that they intend to do so. That represents over one-half of U.S. reactors. We expect virtually all our nuclear plants will renew their licenses—simply because it makes good economic sense to do so.

With license renewal, our first plants will operate until the 2030s and our newest plants will run past 2050. As an industry, we've implemented systematic programs across the industry to manage the systems and components in these plants for their entire expected lifetime. And we're making the capital investments necessary to allow 60 years of operation at sustained high levels of safety and reliability.

Increasing electricity production at nuclear power plants is a key component of the president's voluntary program to reduce the greenhouse gas intensity of the U.S. economy. In December 2002, NEI responded to President Bush's challenge to the business community to develop voluntary initiatives that would reduce the greenhouse gas (GHG) intensity of the U.S. economy. NEI indicated that the U.S. nuclear energy industry could increase its generating capability by the equivalent of 10,000 MW. NEI's analysis showed that this would achieve approximately 20 percent of the president's goal.

The additional 10,000 MW would come from three sources:

- *Power Uprates*—5,000 to 6,500 MW of capacity additions between 2002 and 2012.
- *Improved Capacity Factors*—the equivalent of 3,000 to 5,000 MW of additional capacity in 2002-2012.
- *Plant Restarts*—refurbishing and restarting Tennessee Valley Authority's Browns Ferry Unit 1 would add 1,250 MW.

The nuclear energy industry has recorded substantial progress toward its goal. The NRC has approved 2,198 MW of uprates in the past several years. In addition, based on information from nuclear plant operators, the NRC expects applications for an additional 1,886 MW of uprates in the 2004-2008 period.[5]

---

[5] All power uprates must be approved by the Nuclear Regulatory Commission. Once NRC approval is received, generating companies schedule power uprates into their ongoing capital investment programs. Typically, it takes at least two to three years from the time of NRC approval before the uprate is completed. Given these lead times, companies in 2003 were com-

Continued

In addition, the Tennessee Valley Authority (TVA) is moving forward with refurbishment of Unit 1 of the Browns Ferry nuclear power plant. The TVA Board in May 2002 approved the refurbishment and restart, a $1.8 billion project, that is expected to return the reactor to commercial operation in 2007. Browns Ferry Unit 1 is not a new construction reactor, but its comprehensive refurbishment and restart, when complete, will represent a significant accomplishment for the industry.

With 5,334 MW of new capacity in prospect (4,084 megawatts of uprates and 1,250 MW at Browns Ferry Unit 1), the nuclear energy industry will be approximately halfway toward meeting its goal of expanding capacity by 10,000 megawatts by 2012. This represents substantial progress the largest progress of any single industry—toward achievement of the president's goal to reduce the GHG intensity of the U.S. economy by 18 percent by 2012.

Obviously, there are limits on how much additional electricity output can be produced at the existing 103 nuclear power plants. Meeting the nation's growing demand for electricity—which will require as much as 400,000 MW by 2025, depending on assumptions about electricity demand growth[6]—will require construction of several new nuclear power plants in the years ahead.

STIMULATING INVESTMENT IN AMERICA'S CRITICAL ENERGY INFRASTRUCTURE, INCLUDING NEW NUCLEAR POWER PLANTS

NEI believes that lack of investment in our nation's critical energy and electric power infrastructure is a major problem. Our country is not investing enough in new baseload coal and nuclear plants, and we are not investing enough in new electricity transmission.

NEI's assessment shows that approximately 183,000 megawatts of electricity generating capacity is 30-40 years old; approximately 104,000 MW is 40-50 years old. That represents about one-third of U.S. installed electric generating capacity, and is clear evidence that we are underinvesting for our energy future—relying too much on old, less efficient generating capacity and not investing in new, more efficient and cleaner facilities.

Investment in our country's electricity transmission system has fallen by $115 million per year for the last 25 years, and investment in this area in 1999 was less than one-half of the level 20 years earlier—despite dramatic increases in the volumes of electricity being moved to market. One analysis[7] shows that simply maintaining transmission adequacy at its current level (which is widely acknowledged to be inadequate) would require a capital investment of $56 billion by 2010, equal to the book value of the existing transmission system.

Given these facts, we strongly encourage the passage of energy policy legislation to provide broad-based stimulus for investment in new energy infrastructure, including new nuclear plant construction, deployment of clean coal technologies, new electricity transmission and other energy sources.

Passage of legislation that provides such investment stimulus is essential if we hope to preserve the diversity of fuels and technologies that represent the core strength of our energy supply and delivery system. That stimulus can come through shorter depreciation periods, investment tax credits and production tax credits, loans or loan guarantees, or research and development support, depending on the conditions and requirements of each energy source. In addition, renewal of the Price-Anderson Act, which provides insurance for the public in the case of a nuclear reactor incident, is a necessary step in paving the way toward new nuclear power plants.

NEI believes that more appropriate tax treatment of energy investment must be a central feature of energy policy legislation. As a general rule, the electric industry suffers from depreciation treatment that may have been appropriate for another era, when regulated companies with stable long-term cash flows had a reasonable assurance of investment recovery through rates. But 15- to 20-year depreciation periods for investments in generation and transmission assets are unacceptable for an industry operating in a competitive commodity market, where cash flows are highly

---

pleting uprates approved by the NRC in 2000 and 2001. The NRC approved 2,198 MW of uprates between 2000 and 2003 (243 MW in 2000, 1,111 MW in 2001, 711 MW in 2002, 133 MW in 2003) and expects licensees to apply for an additional 1,886 MW of uprates in the 2004-2008 period. The 2004-2008 forecast represents only those uprates about which the NRC has been informed; it does not represent the total remaining uprate potential of U.S. nuclear power plants.

[6] Energy Information Administration, *Annual Energy Outlook 2004*, DOE/EIA-0383 (2004)

[7] *Transmission Planning for a Restructuring U.S. Electricity Industry*, Edison Electric Institute, June 2001.

volatile and there is no guarantee of investment recovery. Current depreciation treatment acts like a brake on new capital investment.

Energy policy legislation should also address another significant factor that could inhibit capital investment: Regulatory uncertainty. This uncertainty has a chilling effect on capital formation and capital investment. Regulatory uncertainty and perceived risks over the licensing process for new nuclear power plants could inhibit capital investment in new nuclear facilities. In the coal industry, uncertainty over environmental requirements, including possible future limitations on criteria pollutants and carbon dioxide, has slowed capital investment in new coal-fired generating capacity or in upgrading existing capacity. Public policy must recognize the impact of these uncertainties and develop mechanisms to address them.

NEI believes that policymakers must recognize the risks and uncertainties in our economic and regulatory systems and also recognize that policymakers have a responsibility to establish mechanisms to contain those uncertainties.

In the electricity sector, the last several years demonstrate what happens when the markets are left entirely to their own devices without necessary policy and planning guidance. The sole reason that gas-fired plants constitute more than 90 percent of the generating capacity built during the past five years is that these plants present the lowest investment risk. However, as trends in natural gas prices through 2003 demonstrate, sole reliance on gas for new generating capacity can expose consumers to punishing price volatility. Excessive reliance on natural gas for power generation also increases prices and limits the supply available to other industries that depend on natural gas as a feedstock. This, in turn, has a ripple effect reflected in higher prices in many other sectors.

By themselves, markets have no way of valuing energy security, fuel and technology diversity, or other legitimate public policy "goods." Few new coal-fired or nuclear plants have entered service over the last decade, even though these plants provide the greatest measure of price stability going forward.

The decision to employ nuclear power as a major energy source in countries such as France and Japan was based on energy security. The governments of both countries originally decided the use of nuclear energy would protect their nations' energy supplies from disruptions driven by political instability and protect consumers from price fluctuations resulting from market volatility. Today, France depends on nuclear energy for more than three-quarters of its electricity demand, and Japan for more than one-third.

The governments of France and Japan have committed to the use of nuclear energy as an essential part of their nations' future energy portfolios for reasons of economics as well as energy security. Other nations with reactors under construction, such as South Korea and Taiwan, have cited energy security as an overriding concern in the energy policy decisions of their respective governments. Despite all of the international activity, the U.S. nuclear energy sector remains by far the world's largest, producing 762 billion kilowatt-hours in 2003—more than the nuclear sectors of France and Japan combined.

If we do not employ policy mechanisms and investment stimulus to preserve fuel diversity, we run the risk of placing demands on certain fuels that they may not be able to meet. We must address electricity supply and transmission as an integrated system: More coal and nuclear electricity can reduce supply and price pressure on natural gas. More electricity from nuclear plants and renewable sources can moderate environmental pressures and compliance costs that would otherwise be imposed on coal-fired plants.

CREATING THE BUSINESS CONDITIONS FOR NEW NUCLEAR PLANT CONSTRUCTION

NEI believes that our nation must meet rising electricity demand—50 percent growth by 2025—with a diversified portfolio of fuels and technologies, including nuclear energy.

We are confident that new nuclear plants can compete with other forms of baseload generation. Our cost targets $1,000 to $1,200 per kilowatt in capital cost are clearly competitive with other baseload electricity generating options.

Given this, the nuclear energy industry and the Department of Energy launched a program several years ago that will position the industry to build new nuclear capacity when needed, by creating the business conditions under which companies can order new nuclear plants.

This is a comprehensive program designed to address the business issues and unanswered questions—including licensing and regulatory issues, development of new plant designs, and financing—could be roadblocks to new nuclear plant construction.

There are three distinct and major phases on the road toward new nuclear plant construction:

1. pre-commercial licensing and design
2. construction of the first few new plants
3. sustained investment in significant numbers of new plants.

Pre-Commercial Licensing and Design. The Energy Policy Act of 1992 created a new licensing process under which the industry must apply for all necessary regulatory approvals from the NRC before significant capital is committed. Reactor sites and designs can be approved in advance. And new nuclear plants will receive a single license for construction and operation—not the separate proceedings that created unwarranted delay in the period between construction and operation of today's plants.

This approach should limit the regulatory risks that impacted the construction and licensing of many of our operating plants. With the new process, complete plant designs must be available before construction begins. This process also allows meaningful input from the public and other stakeholders early on, before the plant is built, when such input can influence plant design and licensing issues. This should avoid the costly delays common to the old way of licensing a nuclear plant. Because the old licensing process did not require all the design and engineering to be complete when the construction permit was issued, it often resulted in extensive public hearings and public input after the plant was built and before it was allowed to operate.

The industry is validating this new, unproven licensing process. In 2003, Dominion, Exelon and Entergy began a three-year process for requesting NRC approval for early site permits. This does not mean that these companies are committed to building new nuclear plants at these sites. The program is designed to demonstrate that the untested early site permit process works as intended. If approved, the companies will be "banking" those sites for possible future use.

DOE has also requested proposals to share the cost of demonstrating the process of preparing and obtaining a combined construction/operating license from the NRC. This approach consolidates both of these licenses, thereby eliminating the separate hearings, reviews and proceedings that can dramatically increase costs. Beginning this year, DOE is expected to co-fund the cost of at least two applications. Like early site permits, combined construction/operating licenses can be "banked" for future use.

In addition to these licensing issues, this first phase also involves completing the first-of-a-kind engineering and design work for preferably two advanced reactor designs, and obtaining NRC safety certification for those designs. These new reactors are designed to improve safety and reduce capital cost so that they are competitive with other sources of baseload electricity.

The first, pre-commercial phase is not a trivial investment. It will cost $400 million to $500 million to complete the licensing demonstrations and the first-of-a-kind design and engineering for one reactor design. The industry expects to share that cost equally with the federal government under DOE's Nuclear Power 2010 program. The private sector therefore would commit $200 million to $250 million to the effort, or up to $500 million for two reactor designs. It is critically important, therefore, that the government provides adequate funding for DOE's Nuclear Power 2010.

If the private sector and the federal government do not share the cost of design, engineering and licensing work on new nuclear plants, the first few new nuclear plants built will be more costly than follow-on plants. This is because the first plants, like the first new plants of any new technology, would have first-time design and engineering costs associated with them. The industry estimates the capital cost of the first few nuclear plants built would be in the range of $1,400 per kilowatt. After these plants are built and the first-of-a-kind design and engineering costs have been recovered, subsequent plants of the series will have capital costs in the $1,000-$1,100 per kilowatt range, which is fully competitive with other sources of baseload electricity.

Construction of the First New Nuclear Power Plants. Companies interested in building the first new nuclear power plants must address two major challenges: potential regulatory risks and the significant capital investment associated with the first few new nuclear power plants.

Although industry/government programs are seeking to eliminate uncertainty from the licensing process, there is potential for unanticipated cost increases as a result of delays during construction or delays in commercial operation of a completed plant. These delays could be caused by the NRC's failure to deliver necessary approvals on time, or by court challenges to agency actions that are later dismissed. These regulatory risks are beyond the private sector's control and would jeopardize private sector investment in new nuclear power plants.

The financial community has indicated that it is unlikely to provide external debt financing from the capital markets, given the regulatory risks associated with the first several new nuclear power plants. This means that companies considering building new nuclear plants must either finance the first few plants with 100 percent equity, or obtain government loans, loan guarantees, or some other form of comparable government insurance against potential regulatory risks.

Nuclear power plants, like coal-fired power plants, are capital-intensive projects. A company building a new nuclear power plant will invest between $1.5 billion and $2 billion, including interest, during construction. During construction, a company would be investing substantial amounts of equity capital in the project, and this equity would be tied up for a four-to-five year construction period without generating any return to the company. Raising the equity capital required would dilute shareholders' equity and earnings per share. This could lead to lower stock prices, reducing the company's attractiveness to the financial community.

The $18-per-megawatt-hour production tax credit provided in the conference report for H.R. 6 is an important step toward making investment in the first few new nuclear plants attractive to the private sector. This tax credit is comparable to that provided for other sources of new, emission-free electricity generation. The production tax credit would provide an acceptable return on equity, even to a project financed entirely with equity capital. It does not, however, appear to protect the private sector investment against potential regulatory risk, and the industry is continuing to work with the executive branch and Congress to create the financial mechanisms necessary to do that.

Sustained Investment in Significant Numbers of New Nuclear Plants. Companies building new nuclear power plants face two significant challenges: (1) the earnings dilution during construction resulting from the large equity investment over an extended period in a capital-intensive project, and (2) the fact that substantial capital investment would be at risk for an extended period of time. These financing challenges are not unique to nuclear power plants. In fact, they are common to all capital-intensive elements of the electricity infrastructure, including advanced coal-fired power plants and new electric transmission capacity.

Both problems can be addressed through tax-related incentives. An investment tax credit would mitigate earnings dilution and the resulting negative impact on a company's shareholders.

More appropriate depreciation treatment would address the concern over significant investment exposed over an extended period of time. Under current law, nuclear power plants are treated as 15-year property. This depreciation period may have been appropriate for a regulated, cost-of-service business environment. It is not suitable for a competitive, commodity business environment. More appropriate depreciation schedules—seven years instead of 15—would allow faster recovery of investment through reduced income tax liability. Such updated tax treatment would simply recognize that depreciation conventions established for a regulated, cost-of-service business environment are not appropriate for a competitive, high-risk business environment.

It is important to remember that these three phases comprise an integrated program. The pre-commercial activities (like validating the licensing process) are inextricably linked to the financial incentives and investment stimulus for plant construction. Unless the financial incentives for commercial deployment are in place, the companies and the federal government have little reason to invest hundreds of millions of dollars in design and licensing work. And unless we work together to invest in the design and licensing work, there is little reason to create financial incentives and investment stimulus for new plant construction.

INDUSTRY CONFIDENCE IN THE COMPETITIVENESS OF NEW NUCLEAR POWER PLANTS

The nuclear energy industry has a high level of confidence that new nuclear power plants can be built for an "overnight" capital cost[8] of $1,000-$1,200 per kilowatt of capacity for subsequent plants.[9] At this cost, which can be achieved after the first several new plants have been built, new nuclear power units are fully competitive with other baseload electricity production. The financial stimulus sought from the federal government is intended, in part, to "jump start" construction of the first few new nuclear power plants, thereby allowing the nuclear industry to reach a cost level of $1,000-$1,200 per kilowatt for successive plants of that kind. The

---

[8] "Overnight" capital cost does not include interest during construction and is a standard means of comparing the capital costs of various generating options.

[9] This capital cost is achieved after first-time design and engineering costs have been recovered and as industry incorporates improvements in construction techniques and construction management gained during construction of the first few units.

major alternatives to new nuclear plants include conventional coal-fired power plants with a full suite of environmental controls, which have capital costs in the range of $1,000-$1,500 per kilowatt of capacity. These include the so-called "clean coal" technologies, which have capital costs in the range of $1,200-$1,500 per kilowatt of capacity. At $1,000 per kilowatt, a new nuclear power plant could compete with new combined-cycle, gas-fired power plants, which have capital costs in the range of $600-$700 per kilowatt of capacity. Unlike the nuclear and coal-fired technologies, however, gas-fired power plants are extremely sensitive to fuel prices. Economic analysis shows that a new nuclear unit at $1,000 per kilowatt of capacity is competitive with a new gas-fired combined cycle plant fueled with gas at $4-$5 per million Btu.

The cost estimates for new nuclear power plants reflect a high degree of analytical rigor and are as solid as can be achieved, short of actually building a power plant and totaling the dollars spent. Two new designs—the AP1000 developed by Westinghouse and the Advanced Boiling Water Reactor (ABWR) developed by GE Nuclear Energy—serve as examples.

Westinghouse is currently pursuing NRC design certification of its AP1000 nuclear plant. The AP1000 is a 1,117-megawatt Advanced Light Water Reactor (ALWR). It is essentially a higher-power version of a 600-megawatt design, the AP600, which was certified by the NRC in 1999. More than $400 million was invested in developing and licensing the AP600 design, including an extremely detailed cost database, comprising more than 1,900 commodity categories and 25,000 specific items. The cost estimate was verified by Westinghouse, several international architect-engineers, the EPRI and several utilities. A comparably detailed cost estimate was prepared for the AP1000 by modifying the AP600 estimate to reflect the design changes.

In 2002, an industry team—comprised of Westinghouse, seven major U.S. power companies and architect-engineer Bechtel—completed a $1 million re-evaluation of the AP1000 reactor design. As part of that re-evaluation, Bechtel performed a thorough review of the modifications made to the original cost estimate and, after making minor adjustments, endorsed the AP1000 cost estimate.

Although the specific numbers are proprietary, the overnight capital cost for building the first two AP1000 reactors at one site is less than $1,400 per kilowatt. This includes all the first-time costs for completing design, engineering and licensing of the first project. After the first few projects have been completed, the capital cost for later plants will be approximately $1,000 per kilowatt, which is competitive with other sources of baseload electricity. Once those first reactors are built and capital costs reach the $1,000-per-kilowatt range, all future plants would be financed and built without federal government financial assistance.

The Westinghouse-Bechtel estimate of less than $1,400 per kilowatt has a solid analytical basis, has been peer-reviewed and reflects a rigorous design, engineering and constructability assessment.

GE Nuclear Energy and its partners have built two ABWRs in Japan, and are building two reactors in Taiwan (the Lungmen project). In 2002, GE and Black & Veatch (B&V) completed an independent cost estimate of the ABWR. This study resulted in volumes of data, including quantities, vendor costs and construction labor rates. The source of information for every piece of data is referenced. Most references for quantities of materials are to the Lungmen project database, and thus accurately reflect what would be required to build a plant.

This cost estimate was reviewed by GE, B&V and a U.S. utility. The estimate was based on actual experience from current and previous ABWR projects, and is considered a valid forecast of new reactor costs.

The bottom line: a single unit ABWR could be built for $1,445 per kilowatt. Two units on the same site roughly one year apart would have an average cost of $1,300 per kilowatt. These estimates are for a 1,450-megawatt reactor and include owner costs, supplier profit and contingencies.

These costs are slightly higher than the estimates for the AP1000 because the AP1000 incorporates a number of "passive" safety features that reduce the capital cost. GE Nuclear Energy is developing a boiling-water reactor design that incorporates similar advanced passive safety features. The company expects that overnight capital cost for this design will be lower than for the ABWR.

CONCLUSION

Electricity generated by America's nuclear power plants over the past half century has played a key part in our nation's growth and prosperity. Nuclear energy produces more than 20 percent of the electricity used in the United States today without producing air pollution. As our energy demands continue to grow in years to come, nuclear power should play an even greater role in meeting those needs.

The nuclear energy industry is operating its reactors safely and efficiently. In addition, the industry is striving to produce more electricity from existing plants. The industry is also developing more efficient, next-generation reactors and exploring ways to build them more cost-effectively.

The public sector must help create the conditions that will spur investment in America's energy infrastructure, including new nuclear power plants. The passage of comprehensive energy legislation that addresses the business and regulatory risks of building new plants is an important step. The federal government also must continue to support efforts that encourage the industry to continue pursuing new plants, such as Nuclear Power 2010. Finally, Congress must enact policies that recognize nuclear energy's contributions to meeting our growing energy demands, ensuring our nation's energy security and protecting our environment.

Mr. Chairman, on behalf of NEI, I thank you for the opportunity to discuss nuclear energy's significant role in providing electricity to our nation today and its vital importance as a clean, reliable and safe energy source for the future.

Senator ALEXANDER. Thank you, Mr. Fertel.

Mr. Bernhard.

STATEMENT OF J.M. BERNHARD, JR., CHAIRMAN AND CEO, THE SHAW GROUP, INC.

Mr. BERNHARD. Thank you, Chairman Alexander and Senator Craig, for allowing me the opportunity to testify today. My name is Jim Bernhard. I am the chairman, chief executive officer and founder of The Shaw Group.

The topic for today's hearing, nuclear power generation, is a particularly important one. Providing for the Nation's growing energy demands safely, securely, at reasonable costs and with minimal impact to the environment are challenging goals. Last year nuclear power provided over 20 percent of America's electric supply safely and with near record high capacity factors. The Shaw Group takes great pride in being part of the rejuvenated nuclear industry.

In 2000, The Shaw Group purchased the assets of Stone & Webster. As you already know, Stone & Webster is a recognized leader in the nuclear industry with over 115 years of experience as a premier architect/engineer. Stone & Webster and Westinghouse, for example, were responsible for building the Nation's first commercial nuclear power station at Shippingport, Pennsylvania. Since those early days, Stone & Webster has been involved in the engineering and/or construction of over 17 nuclear power stations. We have provided services in one form or the other to over 95 percent of the operating plants in America.

Shaw is already a major player in the environmental infrastructure business. Although you may not be aware of it, Shaw has worked right here in the Senate office buildings, providing the anthrax cleanup at the Hart Building next door. Shaw is active throughout the Government sector, supporting infrastructure at our Nation's military bases, homeland security efforts, various nuclear and non-nuclear cleanup activities, and working closely with the national laboratories and the Corps of Engineers.

I would like to take this opportunity to touch upon three key aspects. First, the Browns Ferry restart. The Tennessee Valley Authority is a long-term and valued client of The Shaw Group. Shaw has supported TVA at all of its nuclear units for over 3 years and we currently provide maintenance and modification support to TVA's operating units.

In 2002, TVA made a decision to restart Browns Ferry unit 1. This $1.8 billion effort is slated for completion in the spring of 2007, and The Shaw Group is providing the construction services for this very important project.

Shaw currently has over 1,300 staff working at the Browns Ferry restart, and we expect that number to peak at roughly 1,700. Most of the staff are local hires, greatly supporting the economics of northern Alabama and Tennessee.

Our scope of work for the project includes asbestos abatement, retubing the main condenser, extensive control room panel and instrumentation upgrades and modifications. Over 35,000 feet of pipe is being replaced, along with approximately 674,000 feet of electrical cable. To date the restart effort at Browns Ferry unit 1 is on schedule and within budget.

Worker safety on the project is our top priority. Our incident and lost time rates are some of the best in the industry and, I am happy to report, is far below the industry averages.

Number two, our engineering technological advances and engineering construction methods have not stood still since the last large plant was constructed in the United States. The status quo is not an option in a highly competitive and dynamic nuclear service market. Technological advances such as modularization, induction in cold bending the pipe, automatic welding techniques to name a few.

Improvements in computer software and hardware have been put to good use. Today Stone & Webster is involved in design of two 1,350 megawatt advanced boiling water reactors on the island of Taiwan. Before a single piece of concrete, steel, or piping is put in place, our engineers and designers built the turbine hall in three-dimensional virtual reality using sophisticated computer-aided design software. Today's electronic methods offer us far greater coordination, communication, and control of these projects. These advances have not only served to reduce the costs and duration of the project, but further improve the project's overall quality and reliability.

Nuclear activities, number three. Progress on the nuclear front is widespread. For example, nuclear power uprates have already added over 2,000 megawatts to the grid by 2000 and should result in another 1,500 by the year 2007. Stone & Webster has been in the forefront of the nuclear power uprates in this country, with involvement in over 38 different sites.

The nuclear industry is also working hard to develop the next generation of reactor design, generation IV.

Recently the Idaho National Laboratory has been designated the Department of Energy's center for expanded nuclear energy activities. We are excited about DOE's plan to rebuild a demonstration generation IV reactor at INL and to couple that with hydrogen generation. Just as the race to the moon in the 1960's resulted in a myriad of offshoot technologies that we rely on today, it is these type of grand visions that will become the basis of our future energy infrastructure.

On the regulatory front, steps have been taken to reduce the license risks associated with new nuclear construction.

Maintaining a trained nuclear workforce is vitally important.

The Congress has done much in the past year to support nuclear research and development throughout the yearly appropriations bills. These modest expenditures have reinvigorated the research community and spurred increasing enrollments at our Nation's nuclear universities. As a company who already employs thousands with nuclear backgrounds, we need this new supply of talented recruits to work hand in hand with our experienced staff to become the next generation of our nuclear professionals.

In closing, with growing concerns over climate change, nuclear power is a source of reliable baseload power that avoid greenhouse gases and other atmospheric pollutants. Nuclear power's contribution to the Nation's energy supply and security deserves recognition. We must redouble our efforts to ensure that the prospect of plant construction in the not too distant future becomes a reality, a reality that provides greater energy independence and energy security for our great Nation.

Members of this committee have been instrumental in the development and stewardship of the energy bill as it currently stands, and I thank and commend you on the work so far. I urge you to continue efforts in that respect.

Mr. Chairman, I would like to thank you for the opportunity to present my testimony today.

[The prepared statement of Mr. Bernhard follows:]

PREPARED STATEMENT OF J.M. BERNHARD, JR., CHAIRMAN AND CEO, THE SHAW GROUP, INC.

INTRODUCTION

Thank you, Chairman Alexander, Ranking Member Graham, and members of the committee and staff for affording me the opportunity to testify before you today. I am truly honored to be here.

My name is Jim Bernhard and I am the CEO, Chairman of the Board, and founder of The Shaw Group. With The Shaw Group corporate headquarters, and indeed my home of many years, located in Baton Rouge, I would particularly like to thank the Senator from the great State of Louisiana, Ms. Landrieu.

The topic for today's hearing—Nuclear Power Generation—is one that is particularly important to me, the Shaw Group, and indeed the nation as a whole. Providing for the nation's growing energy demands, safely, securely, at reasonable cost and with minimal impact to the environment is a challenging goal. Luckily, the nation's 103 operating nuclear power plants are up to the task and performing at record levels. Just last year nuclear power provided over 20% of America's electric supply or an estimated 762 billion kilowatt-hours of electricity. It did so with exceptional levels of safety and security and with near-record high capacity factors approaching 90%. The Shaw Group and its Stone & Webster subsidiary have played a leading role in helping the industry obtain those impressive numbers.

I founded The Shaw Group in 1987 in Baton Rouge as a small pipe fabricator dedicated to supporting the power and process industries. Through the years, The Shaw Group has grown, internally and through strategic acquisitions, to become the nation's number one supplier of fabricated piping to the power, process and petrochemical industries. Shaw fabricates specialty alloy and standard carbon and stainless steel piping, fittings and pipe supports used throughout modern nuclear and fossil power plants. Our nine domestic shops and four international fabrication shops currently have the capability to produce an aggregate of 42,000 pipe spools (9,000 tons of product) per month using the latest manufacturing and pipe bending technology.

In 2000, The Shaw Group purchased the assets of Stone & Webster. As you hopefully already know, Stone & Webster is a recognized leader in the nuclear industry with over 115 years of experience as a premier Architect-Engineer to the power industry. Stone & Webster and Westinghouse were responsible for building the nation's first commercial nuclear power station at Shippingport, Pennsylvania and Stone & Webster engineers were actively engaged in the Manhattan Project and the building of Oak Ridge National Laboratory. Since those early days, Stone & Webster has been involved in the engineering design or construction of over 17 nuclear power stations. We have provided services, in one form or another, to over 95% of the operating plants in America.

Following the Stone & Webster acquisition, Shaw acquired the former "IT Group" in 2002. This acquisition brought to Shaw consulting, construction and technology leaders in the environmental and government services arena. The IT Group was merged with other Shaw assets to form our Environmental and Infrastructure (E&I) subsidiary. Although you might not be aware of it, Shaw E&I has worked right here in the Senate office buildings, providing the anthrax cleanup services in the Hart Building next door as well as the extensive cleanup at the local Brentwood postal facility. Shaw is active throughout the government sector—supporting infrastructure at the nation's military bases, homeland security efforts, various nuclear and non-nuclear cleanup activities such as FUSRAP and Chemical Demilitarization, and is working closely with the national laboratories and Corps of Engineers. We are actively supporting the rebuilding of Iraq and have recently opened an office in Baghdad.

Currently, the integrated Shaw Group employs over 15,000 employees worldwide. Our revenue in 2003 was over $3.3 billion, the majority of which, 85%, came from US operations. Power generation represented roughly 50% of our revenue for 2003 and comprises 30% of our future workload. Nuclear power-related activities were $1.1 billion of last year's revenue. Needless to say, nuclear is a big part of The Shaw Group. I take great pride in Shaw being the world's only vertically-integrated supplier of services to the power industry. Our services can take a plant from cradle to grave with permitting, engineering and design, fabrication of piping and steel, construction, commissioning, operation and maintenance and eventually decommissioning. The Shaw Group has achieved unprecedented growth throughout our history due in large part to technical innovation and an entrepreneurial culture.

Shaw remains committed to the nuclear industry. One of the challenges facing the nuclear industry today is maintaining a highly-skilled nuclear workforce. Through our varied involvement in nuclear projects, Shaw has been able to maintain, and indeed grow, a highly experienced cadre of engineers, designers and craft labor who support the industry. We are one of only a few who maintain the American Society of Mechanical Engineering (ASME) Section III N stamp qualifications and we are members of both the Nuclear Energy Institute (NEI) and the Institute of Nuclear Power Operations (INPO).

The nuclear industry today is more robust than it has been in many years. Part of that is the result of the exceptional high performance and safety levels attained by the operating nuclear fleet. As I mentioned, plant capacity factors have risen to nearly 90%, with refueling and maintenance outages being completed in record times. Nuclear-generated electricity is the low cost leader at an average cost of generation of only 1.7 cents/kilowatt-hour.

Credit for the recent nuclear revitalization is to be shared among many parties. The utilities have made great strides at improving operations and maintaining their plants. NEI has also done much over recent years to support its member utilities and deserves credit. Furthermore, the NRC is to be acknowledged for streamlining its review process while maintaining the firm oversight role that is the heart of its mission.

The Shaw Group is also part of the nuclear success story. As the nation's largest provider of craft labor to support maintenance and modification efforts at the plants—peaking at over 5000 employees—we have striven for and achieved great success in helping our clients meet their performance and safety goals.

Having provided that short background on The Shaw Group, I would like to take the opportunity today to touch upon three key aspects. First, I will address Shaw's role in the restart of TVA's Browns Ferry Unit 1 which I believe is of interest to the Chairman and committee. Second, I would like to mention some of the technological advances that have taken place since the nation's last nuclear plant was built and the role of Architect-Engineers like Shaw in bringing those advances to bear to help revitalize the nuclear industry. Finally, I wish to hit upon other developments in the nuclear arena and some of the challenges, and more importantly, the opportunities that we face as we move forward with a revitalized nuclear power sector.

RESTART OF BROWNS FERRY UNIT 1 AND SUPPORTING THE TVA FLEET

Serving over 8.3 million customers in the Southeast, the Tennessee Valley Authority is a long-term and valued client of The Shaw Group. With five operating reactors at three locations providing over 5700 MWe of power, TVA plays a key role in the U.S. nuclear industry. Stone & Webster has provided maintenance and modification services to TVA for the past seven years and will continue in that capacity under a recently renewed contract. Just last week, Shaw had nearly 750 staff supporting TVA maintenance and modifications efforts at their Sequoyah, Watts Bar and Browns Ferry operating plants.

In 2002, TVA made the decision to restart the Browns Ferry Unit 1 boiling water reactor which had been shutdown since March of 1985. The $1.8 billion effort is slated for completion in the spring of 2007. The Shaw Group is providing the construction and construction management services for the restart. We are uniquely suited to the task, having successfully provided similar efforts during the Browns Ferry Unit 3 restart effort that ended in 1995.

Shaw currently has over 1300 staff working on the Browns Ferry 1 restart effort in the field. We expect the maximum number of staff to approach 1700 at the peak of the forecasted work. The majority of staff (80%) is local hire, greatly supporting the economy and small businesses of northern Alabama and Tennessee. Of all subcontracted work awarded by Stone & Webster Construction Inc. in the last quarter of 2003, 71% was awarded to local Tennessee Valley firms, 79% was awarded to small businesses and 49% was awarded to Disadvantaged, Woman-Owned, or Veteran-Owned Businesses.

Our construction restart services include asbestos abatement, re-tubing the main condenser, support activities to refurbish the main turbine, turbine generator and associated pumps and valves along with extensive control room panel and instrumentation upgrades and modifications Over 35,000 feet of large and small bore pipe is being replaced along with 674,000 feet of electrical cable and 142,000 feet of electrical cable tray and conduit. To date, the restart effort at Browns Ferry Unit 1 is on schedule and within budget, with the asbestos abatement and condenser re-tubing complete, and over a third of the large bore piping already installed.

I fully expect the Browns Ferry Unit 1 restart project to be a success and bring online much-needed additional generation capacity to the growing Southeastern US economy. I should clarify—to a very large extent "emission free" additional generation, since nuclear power plants avoid millions of tons of nitrous and sulfur dioxides and carbon dioxide from being introduced into the atmosphere.

Make no mistake about it, the Browns Ferry Unit 1 restart program is a large effort, and, any project of this magnitude and scope faces risks which are, to a large extent, outside the immediate control of the project team. Labor unrest, material shortages, and extreme commodity price swings, are examples of risks that have the potential for delaying a project. This far into the project, the availability of material and labor has had positive impacts on both schedule and cost. And, it is our mission to bring about the restart safely, with high quality and reliability, on time and on budget. The BF-1 restart team fully expects that these goals will be met.

One aspect of the Browns Ferry 1 restart is the attention to safety of those working on the project. Shaw prides itself on its safety program and it is an integral part of every employee's job through our ShawSAFE program. The Shaw Group incident and lost-day rates are some of the best in our industry, far below industry averages, and we have won numerous local, state and nationwide awards for our commitment to safety. We will continue to work closely with TVA, and indeed all of our utility clients, to maintain our excellent safety record.

NEW TECHNOLOGIES, INDUSTRY DEVELOPMENTS

TVA's Watts Bar Unit 1 was the last plant brought into commercial operation in the United States in 1996. Since that date, no new plants have been brought online in the US and no new plants have been ordered since 1979. However, that doesn't mean that the nuclear engineering and construction industry has stood still. It hasn't.

The nuclear engineering and construction business continues to incorporate new technology into the way it does business. Standing still in this respect is not an option—otherwise your competitors will overtake you. And let me assure you that the nuclear services industry is a highly competitive and dynamic marketplace.

A large part of any nuclear plant is "piping". I mentioned earlier the manufacturing and bending technologies that Shaw utilizes. Our induction bending machines have the ability to bend a 66 inch diameter pipe with as much as a 5 inch wall thickness. The importance of induction or cold bending is the fact that this reduces welding in the field thereby reducing both cost and schedule on large projects. Advances in automatic or "orbital" welding have led to greater quality of critical piping welds along with reduced time and cost. Bar-coding of piping and other equipment allow for ease of tracking during manufacturing and construction. Advances in engineering and construction technology not only serve to reduce the cost and duration of a project but further improve the project's overall quality and reliability.

Progress in the computer sector has likewise led to significant advances in engineering and design. Today, Stone & Webster is involved in the design of two 1350 MWe Advanced Boiling Water Reactors (ABWR) on the island of Taiwan. Before a single piece of concrete, steel or piping is put in place, our engineers built the turbine hall in three-dimensional virtual reality using sophisticated Computer Aided Design software. This ensures that everything fits together properly without interferences, further reducing construction efforts in the field. The software generates the design drawings from which the plant piping is manufactured and ultimately from which the entire plant is constructed. Furthermore, today's designs are "intelligent" in that critical information from specifications, calculations and drawings are maintained and shared across databases affording us far greater access to information for tracking and control over the duration of the project. Today there is far greater coordination between the engineer, manufacturer, constructor and ultimately operator of power plants. The computer tools of today provide the information needed to better manage large projects. Needless to say, communications technologies have greatly improved in recent years and they are being adeptly put to use.

Another significant advance is that of "modularization." Assembling the various parts of a large nuclear plant is a complex technical and logistical challenge. Any time that portions of the plant can be pre-assembled in a controlled environment and then shipped to the construction site for installation, cost and schedule savings are achieved. Shaw has extensive experience in modularization, with a dedicated 60 acre Gulf-Coast facility used for both power and petrochemical projects.

Shaw and the TVA Browns Ferry 1 restart team are bringing numerous technical advances to bear. Gamma scanning equipment has been used to locate high radiation sources so that we can eliminate them and lower overall radiation exposure. To the greatest extent possible we modularize our equipment purchases. The project utilizes special machining equipment to cut and prepare piping spools in the field along with state-of-the-art automatic welding equipment for critical piping welds. Ground penetrating radar equipment is used to locate rebar in the concrete walls and slabs prior to installing concrete anchors. Laser templating has been used to record as constructed locations and laser photogrammetry will be employed for critical piping needs. Three dimensional computer modeling and graphics are employed to display the design configuration of components within the containment drywell and further serves to assist in planning the work sequences necessary to implement plant modifications.

As we all know, the cost overruns and schedule delays of some plants built in the 1970's and 1980's are entirely unacceptable in today's marketplace as they were then. The future of nuclear power in the U.S. demands that any new project be completed on schedule and under acceptable capital risk. Current expectations in the industry are that a large grassroots nuclear plant could be completed in less than 48 months and for overnight capital costs of under $1500 per installed kilowatt. All of the technological advances and innovations achieved over the past years, and an experienced and trained nuclear workforce, in both engineering and construction, must be brought to bear to meet those schedule and cost goals. There is no single entity that can bring about the success of a new nuclear plant. It will require a dedicated team of utility, reactor vendor (NSSS), equipment manufacturers, and architect-engineer firms all working together with a common goal.

CURRENT NUCLEAR ACTIVITIES AND THE ROLE OF THE ARCHITECT-ENGINEER

Progress on the nuclear front is widespread; more power is being generated from our existing stations, new reactor designs are being developed, the co-generation of hydrogen from nuclear power is gaining momentum, new nuclear plants are being completed internationally, and the regulatory framework for new nuclear construction domestically is in place. Importantly, enrollments at the nation's nuclear engineering programs are beginning to grow. A revitalized industry is poised to meet the nation's energy and security demands.

Approved nuclear power uprates, for example, have already added 2035 MWe to the grid since 2000. Another 240 MWe are under review and over 1286 MWe in uprates are planned for implementation by 2007. That is the equivalent of adding three new large nuclear stations. Stone & Webster is at the forefront of nuclear power uprates with involvement at 38 units covering 24 different stations. Power uprates, reduced refueling outages and better performance, combined with restart of shuttered units such as Browns Ferry 1, are resulting in ever more megawatts being delivered from the already-licensed existing plants.

Current efforts are now underway to develop the next generation of reactor design or Generation IV plant. Such a design will address the challenging goals of being highly economical, with efficiency levels comparable to that of combined cycle natural gas plants, minimum waste generation, walk-away safety features and proliferation resistant fuel.

Closely coupled to the Generation-IV effort is the growing technology related to hydrogen production. Fuel cells are rapidly gaining greater acceptance, both for stationary as well as transportation applications. New developments are taking place rapidly on this front. At their heart, fuel cells require hydrogen, either in pure form or from reformed hydrocarbon stocks. The infrastructure changes implicit with greater hydrogen use are no small challenge. New nuclear designs, particularly the emerging high temperature gas cooled reactor designs, hold great potential for being capable of not only producing electricity but also being co-located with a hydrogen production facility.

Recently, the Idaho National Laboratory has been designated as the Department of Energy's epicenter for expanded nuclear energy activities. The Idaho lab has its roots in applying technology to meet the needs of society and, in particular, its energy needs. We are keenly interested in the success of the new INL, and indeed all of our national labs, since it is there and in the universities where tomorrow's technology is developed. We are all very excited about the DOE's plans to build a demonstration Generation-IV reactor at INL and to couple that with hydrogen generation. Just as the race to the moon in the 1960's resulted in a myriad of offshoot technologies that we rely on today, it is this type of grand vision that will become the basis for our future energy infrastructure.

Also in the government sector, the design of the Mixed Oxide Fuel Fabrication facility at the Savannah River Site continues with the goal of turning the nation's stockpile of surplus plutonium to usable fuel and thus electricity. Shaw continues to be a leader in this area through the Duke-Cogema-Stone & Webster consortium.

Progress is not only limited to the United States. New construction of nuclear plants continues in other parts of the world and many nations obtain an even greater percentage of their energy supply from nuclear than we do. Finland has recently decided to build a new nuclear station its fifth. New construction in China, South Korea, and Taiwan continues. In Taiwan, Shaw is actively engaged in completing the design of the balance of plant for the two Lungmen ABWR plants. We also are providing consulting engineering services for the four new units, Shin Kori 1&2 and Shin Wolsong 1&2, in South Korea.

On the regulatory front, steps have been taken to reduce the capital and licensing risk aspects of new nuclear construction. The new "certified designs," either those already approved by the NRC or those currently under review, will greatly reduce licensing uncertainty. The combined construction/operating license approach under 10 CFR 52 will also help and the DOE is requesting cost-share proposals on that front. As you likely know, three nuclear utilities have already applied for Early Site Permits in the hopes of banking a site for possible future nuclear expansion. All of these activities should serve to lessen the risk and enhance the potential for necessary new construction in the U.S.

The Congress has done much in past years to support Research and Development at our laboratories and universities through the yearly appropriations bills. This is vitally important to ensure the future of the nuclear workforce. These modest expenditures have re-invigorated the research community in recent years. As a company who already employs thousands with nuclear backgrounds, we will need this new supply of recruits to work hand-in-hand with our experienced staff and to become the next generation of nuclear professionals.

CLOSING REMARKS

The role of the nuclear Architect-Engineer and Engineer-Procure-Construct contractor is a demanding one. While the researchers and scientists develop the concepts, it is the AE's task to bring those concepts to physical reality. The detailed and exacting design of the plant and its subsequent equipment manufacture and construction is our role and it is a vital one—absolutely necessary to bring about the success of any nuclear facility project. We are excited about developments on the Generation IV designs and we look to the success of DOE's Nuclear Power 2010 initiative. Shaw remains ready to support new nuclear development and construction and is uniquely qualified to do so.

It is important to realize the significance of nuclear power to our ever-growing energy supply. Again, over 20% of our electricity comes from this fuel. The restart of Browns Ferry Unit 1, nuclear power uprates and upgrades at existing plants, and the not-too-distant prospect of new construction all point towards greater energy independence and energy security for the nation. With growing concerns over climate change, nuclear power is one source of large baseload power that does not introduce greenhouse gases and other atmospheric pollutants into the atmosphere during operations.

Although the road ahead looks bright, challenges and obstacles remain. It is critical that an energy bill with the right mix of incentives to promote new nuclear construction is soon passed. The nation needs a comprehensive energy plan one that clearly addresses the importance and needs of the nuclear power sector. Members of this committee have been instrumental in the development and stewardship of the energy bill as it currently stands and I thank and commend you on your work so far. I urge your continued efforts in that respect.

Mr. Chairman, I would like to thank you and the committee for the opportunity to speak here today, and would be happy to answer any questions you may have.

Senator ALEXANDER. Thank you very much, Mr. Bernhard.

Mr. Asselstine.

STATEMENT OF JAMES K. ASSELSTINE, MANAGING DIRECTOR, LEHMAN BROTHERS, INC.

Mr. ASSELSTINE. Thank you, Mr. Chairman.

As I see it from a financial perspective, there are seven requirements that will have to be met if the industry is to be in a position to make commitments to new nuclear power plants in this country and if the financial community is to be comfortable with those commitments. I will summarize those very briefly and then we can move to questions.

The first requirement is the continued strong regulatory and economic performance of the existing plants. We are well through the process of moving to competitive marketplaces in about half of the States in the country and companies that own nuclear assets have fared well in that process. Companies have been given a fair opportunity to recover their stranded costs. Decommissioning costs are viewed as an appropriate expense recoverable from retail ratepayers. And we have seen a fair amount of consolidation within the industry, which quite frankly, I think has contributed to improved performance. We have seen significant improvement in both the reliability and regulatory performance of the plants as well over the past decade. Capacity factors have improved substantially. Reportable events to the NRC have declined significantly. Refueling outages are considerably shorter, and production costs have come down. As a consequence, nuclear units today are very cost competitive from a production cost standpoint with power being generated by gas- and coal-fired power plants.

In addition, as other speakers have already described, we have a new and, I think, improved regulatory framework in place, recognizing the improved performance that we have seen from the plants.

So we have a regulatory framework that supports a move to a competitive industry and significant improvement in the operating performance of the plants.

My second requirement for future commitments to new nuclear plants is that those units will have to be cost competitive with other generation alternatives. At the end of the day, the decision on which alternative to choose will be an economic one. Nuclear and, for that matter, coal-fired plants do face something of a disadvantage in that they are more complex machines. Initial construction costs are higher. Construction periods are longer, and it is no surprise that as you look at your chart, we have seen significant commitments to gas-fired power plants over the past several years which have a lower initial capital cost and can be built during a shorter period of time.

I tend to agree with the industry representatives, that if we can bring in new nuclear plant designs in the $1,000 to $1,200 per kw range, nuclear can and should be very competitive with other alternatives in the future.

My third requirement is the need for a high degree of assurance that a new nuclear unit will be built at a predictable cost and on a dependable schedule. This really turns on two issues: certainty around the construction process and costs and, second, certainty around the regulatory process. One distinction that nuclear has, compared with other alternatives, is the regulatory process that applies to licensing, building, and ultimately operating a new nuclear unit.

We have a new licensing and regulatory framework that was established under the Energy Policy Act and that has been described by other speakers, but that process is, as yet, untested. We are working through tests of the early site permit and design approval portions of the process, but the key ingredient here is the issuance of a combined construction and operating license that should reduce uncertainty in terms of potential for delays in bringing a plant into operation after a substantial amount of capital has been invested in that unit. And that uncertainty will continue until we have tested the process out with a few plants.

My fourth requirement is the need for appropriate financing arrangements to cover the construction costs of a new nuclear plant. There are a variety of ways to tackle the financing of a nuclear unit. Nuclear units, in my view, could be financed, as they have been historically, by a regulated utility using the full assets and cash flows of an ongoing operating utility.

However, given the move to competitive markets, it is unlikely that a traditional regulated utility will be building new generation in the future. It is more likely that plants will be built by competitive generation companies. Clearly investors have been comfortable with several existing competitive generation companies that include operating nuclear units as a substantial part of their generating portfolio. The key is get through the construction process, get the plant in operation, demonstrate that the plant is meeting its performance requirements. After that, I think investors will be comfortable in taking nuclear operating risks, and that is really the key issue that will need to be addressed in the financing arrangements.

Mr. Chairman, my remaining elements are fairly straightforward: public confidence, which will turn on the continued safe and reliable operation of the existing plants, and progress in dealing with the spent fuel disposal problem, and renewal of the Price-Anderson Act to extend the insurance indemnification provisions to new plants as well as the existing ones.

Thank you.

[The prepared statement of Mr. Asselstine follows:]

PREPARED STATEMENT OF JAMES K. ASSELSTINE, MANAGING DIRECTOR, LEHMAN BROTHERS, INC.

Chairman Alexander, Ranking Member Graham, and members of the Subcommittee, my name is Jim Asselstine. I am a Managing Director at Lehman Brothers, where I am the senior fixed income research analyst responsible for covering the electric utility and power sector. In that capacity, I provide fixed income research coverage for more than 100 U.S. electric utility companies, power generators, and power projects. As a research analyst, I also work closely with the large institutional investors who have traditionally been a principal source of debt financing for the power industry. I appreciate your invitation to testify at today's hearing regarding new nuclear power generation in the United States. In my testimony today, I intend to discuss seven requirements that I believe must be met if the industry is to decide to enter into commitments to build new nuclear power plants in this country, and if analysts and investors are to support that decision.

The first requirement is the continued strong regulatory and economic performance of our existing nuclear plants. By way of background, we currently have 103 operating nuclear units in the United States. These units are located in 31 states and are operated by 27 different companies. Together, these plants represent about 97 gigawatts of generating capacity, or about 12 percent of total U.S. capacity. Because these are baseload plants that operate with high reliability, these units produce more than 20 percent of total U.S. electric output. The plants consist of two reactor types: 69 are pressurized water reactors; and 34 are boiling water reactors. Of our existing fleet, the last unit to enter commercial operation was TVA's Watts Bar 1 unit in June 1996.

Following the enactment of the Energy Policy Act of 1992, analysts and investors focused considerable attention on the transition arrangements as we moved from regulated to competitive markets, and especially on the ability of the utilities to recover their stranded costs. (Stranded costs represent the difference between the book value of the utility's assets and their market value in the competitive market.) In many instances, capital investment in the existing nuclear plants represented a substantial portion of the utility's stranded costs. To date, about half of the states have adopted restructuring plans for the power industry. In essentially all cases, these plans have provided the utilities a fair opportunity over the transition period to competitive markets to recover most or all of their stranded costs. Further, the states have provided for the continued recovery and collection of nuclear plant decommissioning costs from retail ratepayers, recognizing that nuclear plant decommissioning is a health and safety requirement and a financial obligation that was largely incurred during the period of regulated operations. We have also seen considerable consolidation in the ownership and operation of the U.S. nuclear plant fleet. This consolidation has taken place through traditional mergers, purchases of nuclear units by other utilities, corporate restructurings, and new operating arrangements. Taken together, these industry restructuring arrangements have treated the existing nuclear plants in a fairly benign manner.

We have also seen significant improvement in the regulatory, operating, and economic performance of the existing plants over the past decade. The number of significant events reported to the Nuclear Regulatory Commission has declined substantially, as has the average duration of refueling outages. Average capacity factors for the U.S. nuclear fleet have improved significantly, and production costs have declined. As a consequence, a well-run single nuclear unit now has production costs, including fuel, operations and maintenance expenses, ongoing capital requirements, general and administrative expenses, and taxes, of about $20/megawatt-hour, and large, multi-unit plants have production costs of below $20/megawatt-hour. These production costs compare very favorably with other forms of generation, including coal-fired and gas-fired power plants. With the current high natural gas price environment, nuclear units, like coal-fired plants, are viewed by both the industry, and analysts and investors, as attractive assets. One issue affecting analyst and investor perceptions of the performance of the existing nuclear plants is the need for effective inspection and maintenance practices to maintain the material condition of the plants. As a result of the extended shutdown of FirstEnergy's Davis-Besse plant, the financial community is sensitized to the adverse economic impacts of poor maintenance practices that result in a substantial degradation of the physical condition of important plant equipment. The industry will need to continue to pursue aggressive inspection and maintenance programs to ensure that material condition problems are identified and corrected at an early stage, before they result in serious degradation of important safety equipment.

My second requirement for future commitments to new nuclear plants is that those units must be cost competitive with other generation alternatives, most notably gas-fired and coal-fired generation. As we move to more competitive power markets, industry decisions on new generation, and how the financial community perceives those decisions, will be driven by the relative cost, and the risks and uncertainties associated with the available alternatives. As discussed above, the strong operating performance of the existing plants demonstrates that production costs for a new nuclear plant should be very competitive with other alternatives, especially if the new plant design represents an evolutionary step beyond the existing plant designs. The other variable is the capital cost of building the plant. Here, new nuclear units, and for that matter, new coal plants, face some challenges when compared with gas-fired generation. Nuclear and coal plants have a more complex construction process, and take considerably longer to build, than gas-fired plants. This results in higher capital costs and higher interest costs during the construction period. Also, a longer time period is required to recover the investment after the plant has entered commercial operation. Taking into account these factors, I agree with the industry representatives that a new nuclear plant will need to have a capital cost in the range of $1,000-$1,200/kilowatt in order to be cost-competitive with the other available alternatives.

My third requirement is the need for a high degree of assurance that a new nuclear unit will be built at a predictable cost and on a dependable schedule. The industry and the financial community remember that a number of the existing plants that received their operating licenses in the 1980s and 1990s experienced delays due to regulatory or licensing issues that arose after most or all of the capital investment in the plant had been made. These delays were caused by a number of factors, including construction issues, quality assurance weaknesses, coordination issues between plant design and construction work, changing requirements, and the mechanics of the two-stage licensing process, which resulted in litigation at the pre-operation stage. The Energy Policy Act of 1992 and subsequent actions by the NRC have put in place a new regulatory process that should result in the resolution of licensing issues at an early stage in the process before large capital commitments to build the plant have been made. This new regulatory process provides for the pre-approval of new, standardized plant designs, allowing for the resolution of regulatory issues and the completion of substantial design work before construction work begins. The process also provides for the pre-approval of nuclear plant sites. As is the case with the design approval process, the use of early site permits should allow major siting questions to be resolved before a decision is made to proceed with a new plant. Finally, and perhaps most importantly, the new process provides for the issuance of a combined construction and operating license. The objective of the combined license, together with an agreement on the regulatory standards to be applied by the NRC in monitoring the construction process, is to resolve all key safety and regulatory issues before the start of plant construction, and to minimize the risk of delays in plant operation after the capital investment has been made. The NRC and the industry are now implementing and validating the standard design approval and early site permit features. This will provide some assurance that the new regulatory process will work as intended. Unfortunately, however, some uncertainty will remain until the first few plants have successfully completed the entire process of receiving a combined license, completing construction, and entering commercial operation. Until we gain this experience for the initial plants, both the industry and the financial community are likely to require some added measures to mitigate this construction completion and initial plant performance risk.

My fourth requirement is the need for appropriate financing arrangements to cover the construction costs of a new nuclear plant. Historically, our existing nuclear units were financed by electric utilities as part of their regulated utility operations. Typically, the utility would demonstrate that the new nuclear unit was needed and represented the best available alternative. Following state regulatory approval and receipt of a construction permit from the NRC, the utility would proceed with construction. Most construction costs were met by the utility with a combination of cash from its other utility operations, and the proceeds of new debt and equity issuance by the utility or its parent company. Recovery of most of the investment in the plant would not take place until after the plant had received an operating license from the NRC, the plant had entered commercial operation, and the state regulators had determined that the investment in the plant was prudent and recoverable from ratepayers. Although there were some unpleasant surprises in terms of state regulatory disallowances of some investments in the current generation of nuclear units, this system worked fairly effectively as a means to finance new plant construction in the 1980s and 1990s. Going forward, a utility that elected to build a new nuclear unit could finance that plant as part of its regulated utility operations.

Given the move to deregulated power markets, however, it is perhaps more likely that a future nuclear unit would be built and operated by a competitive generation company. Investors have been willing to invest in generation companies that have a substantial component of operating nuclear plants in their generation mix, especially if those plants have a solid track record of operating performance, are cost-competitive in their regional markets, and the generation company has stable revenues tied ultimately to retail customers or load-serving entities. Although it would be challenging, it is conceivable that a large competitive generating company with a diverse portfolio of operating assets, could finance the construction of a new nuclear unit with appropriate mitigation of construction completion and initial operation risk. Another alternative would be to finance a new nuclear unit through a consortium of a number of experienced nuclear companies, including utilities or generation companies, and manufacturers and suppliers. The consortium approach has the advantage of limiting the financial risk to any single party, but has other potential operational disadvantages. The most challenging alternative would be to attempt to finance a future nuclear plant on a stand-alone basis without recourse to another company or companies with other assets and revenues. Given the uncertainties associated with an untested licensing process, the length of the construction process, and the cost of the project, this non-recourse financing approach does not appear to be feasible without substantial financial risk mitigation features.

My fifth requirement is the need for a continued low cost supply of fuel and enrichment services given that low and stable fuel costs are an important component of the cost-competitiveness of nuclear units. With ample supplies of uranium, multiple sources of enrichment services, and new proposals for enrichment providers, this requirement appears to pose limited risk.

My sixth requirement is public acceptance. Public acceptance of new nuclear plant commitments will likely turn on two issues: public perceptions of the safety of nuclear plants; and confidence that we will achieve a workable solution for spent fuel disposal. Public perceptions on the safety issue will likely be determined by the ongoing performance record of our existing plants. Continued progress in developing, licensing, building, and ultimately, operating a waste repository will likely be the determining factor on the spent fuel disposal issue.

Finally, extension of the Price-Anderson Act will be needed to extend the nuclear liability indemnification system to new nuclear plants. It is doubtful that the industry or the financial community would proceed with a new plant commitment without this system in place.

Mr. Chairman, your staff also raised several questions regarding TVA's ongoing program to return Browns Ferry Unit 1 to service, and its implications for future nuclear plant development in this country. The Browns Ferry Unit 1 refurbishment and restart effort is a significant undertaking, with a program that is expected to take up to five years and result in up to 2,400 temporary jobs, and with a cost estimate of $1.7-$1.8 billion. It appears that the Browns Ferry Unit 1 refurbishment effort will be the most extensive effort involving a nuclear plant since the completion of the last round of new plant construction in the mid-1990s. As such, I believe that TVA's experience can be very valuable in building confidence within the industry and within the financial community that the scope of construction work on a new plant can be managed effectively. If TVA and the NRC can work effectively on this project, and if TVA can complete the refurbishment process and return Browns Ferry Unit 1 to service within the projected budget and time schedule, this would represent a positive contribution. But because Browns Ferry Unit 1 has an operating license, this refurbishment process will probably not reduce the uncertainties around the as-yet untested combined construction and operating license process. Conversely, cost overruns and delays could have negative implications depending upon the causes.

The experience that engineering, procurement, and construction contractors have obtained on certain international nuclear power projects is also relevant in terms of building confidence within the U.S. industry and the financial community about a future nuclear plant here. Several of the international projects now underway are likely to be similar to the new standardized designs that would form the basis for a new plant order in the United States. Continued success in completing those international projects on budget and on schedule should provide added confidence in the schedules and cost estimates for new U.S. plants. Again, unfortunately, until we gain actual experience with the new NRC regulatory process, that major area of uncertainty will remain.

In terms of potential financial community investment in the restart of Browns Ferry Unit 1, TVA has stated that it expects to be able to fund the cost of the refurbishment program and still achieve its debt reduction objectives. TVA enjoys strong and exceptionally broad-based investor support for its Power Bonds due to its very high credit quality, its status as a wholly-owned corporation of the U.S. government, its successful and low cost generation and transmission operations, and its rate-setting authority. It appears that TVA will be able to execute and finance its Browns Ferry Unit 1 restart effort as currently contemplated. It is possible that the restart program for that unit or another nuclear unit with an operating license could attract other sources of financing if needed, but the more extensive the effort, and the more it resembles the scope and scale of new plant development, the more the financing constraints and conditions for a new plant, discussed above, will apply. Thank you.

Senator ALEXANDER. Thank you very much. I will begin with questions. I will limit myself to 5 minutes and then we will go to Senator Landrieu, and then we will just go back and forth for a few minutes.

If I may go to the TVA Chairman, Mr. Bernhard said that Browns Ferry is, so far, on schedule, on budget. Mr. Asselstine has said that one of the seven elements to building confidence and to creating an environment in which other nuclear power plants can be built is showing that such plants can be built on time and on schedule.

What do we mean by on time and on schedule? I have read the figures, $1.7 billion in the year 2007. Is that what we are talking about when we say on time and on schedule? How do you measure that and what are the critical flags that you watch for to make sure that you are progressing properly?

Mr. MCCULLOUGH. Mr. Chairman, before the board considered this decision, first of all, it was driven by the need for additional baseload to meet the valley's economy. TVA spent 7 months in a detailed scope of work. From that we got a precise estimate of the cost to complete that scope of work and a comprehensive work plan. We call it the DESEP, a detailed estimate of the scope estimate and plan. TVA did not do this alone. We had some of the best minds in the industry, external industry experts, who analyzed the scope of work that would have to be accomplished, the cost, and got a detailed estimate, and then formulated a detailed work plan.

$1.8 billion was the cost. More precisely, $1.777 billion, and 60 months, a 60-month plan. TVA is 41 percent complete according to that plan, and we are on budget.

Senator ALEXANDER. Thank you, Mr. Chairman.

If you should succeed, as you are today, it looks like TVA might have the opportunity to open a second or even a third nuclear power plant, should you choose to do so. What about the possibility of additional power plants at Watts Bar and Bellefonte? Where do those fit into TVA's strategic plan?

Mr. MCCULLOUGH. TVA's plan to build and develop will be driven by our need to supply the baseload demands in the Tennessee Valley. Our projections are right now that with the successful recovery of unit 1 scheduled for May 2007, we would not need additional baseload generating capacity until about 2014. But as you note, we have valuable capacity at Watts Bar 2. TVA also has valuable capacity for future economic growth in demand that could be brought on line at the Bellefonte site. But again, that decision will be driven by the need to furnish additional baseload for the valley.

Senator ALEXANDER. And at the moment, you feel like you have got capacity to meet the needs until 2014.

Mr. MCCULLOUGH. Yes, sir.

Senator ALEXANDER. Are there any changes in the law or regulations—well, let me not say regulations. Are there any changes in the law that we should consider that would make it easier for you to complete this plant in a safe and efficient manner or to consider moving ahead with a second or third plant at some later time?

Mr. McCULLOUGH. Well, Mr. Chairman, we have confidence in NRC's vigilance to ensure that the highest standards of safety are complied with, and it is a continuous improvement process. I know Dr. Travers and his colleagues at NRC worked with you and other members of the Senate.

I would point out that TVA is supportive of the $18 per megawatt hour production tax credit that is a part of the pending energy bill. We think that that is a responsible incentive that investor-owned utilities could benefit from and could enhance the future of new nuclear technology to meet the Nation's baseload demand going forward.

Senator ALEXANDER. TVA itself could not benefit from that incentive. Is that correct?

Mr. McCULLOUGH. That is correct, Mr. Chairman.

Senator ALEXANDER. Thank you. I have some other questions, but I think we will go to Senator Landrieu now.

Senator LANDRIEU. Thank you, Mr. Chairman.

This is to Mr. Bernhard. If you could just state for the record if there was a new plant ordered to be initiated—and we are hoping for the passage of the energy bill. There are a few things that need to be worked out. But in the event that it is passed, it hopefully will lay the groundwork for revitalization of the industry.

Number one, if you could give us just your sense of how many new plants you think might be brought on line, I mean realistically, based on the need and what your understanding of the industry is.

And, are there architect/engineering firms like yours prepared to bid, win, and execute the projects? Are there any complications that you might want to share with the panel based on the skill level, engineering, et cetera since this industry is, I would not say, dormant, but it has been in a non-revitalized state for some time. So could you just make some comments about that?

Mr. BERNHARD. Sure. Thank you, Senator Landrieu.

I cannot speak for all architect/engineering firms, but I certainly can speak to The Shaw Group and we are certainly capable and ready and willing to perform a successful execution of a new plant as we currently are with Browns Ferry, which in a lot of ways is more difficult than building a new nuclear power plant because existing structures have to be coordinated with new materials, et cetera. We have had a good plan there and are well underway of completing that project on time.

One of the key things is having the quantity and quality of people to build a nuclear power plant. We are the largest company in the United States currently with over 5,000 people in the nuclear business serving the nuclear industry. We are doing engineering for Taiwan for Lungmen power units 1 and 2, and we just were awarded a project to do some preliminary engineering for the Korea Electric. So because of companies and facilities outside the United States, our engineering base has been kept current along with the ability to do uprates for the nuclear power industry.

I think the difference here is today, as in all construction, because of the computer-aided design where we are able to do plants a lot more efficiently, not only on planning and scheduling, but actually have a virtual reality of what the plant would look like, it creates an atmosphere of efficiency and reliability that was not available 25 years ago because of technology, and that technology today has allowed nuclear power facilities to be built in a lot surer, more definitive time frame than had been in the past.

The important aspect of building plants going forward is certainty of government regulations from the beginning, and what has hurt the construction process in the past is when regulations or intervenors in the middle of the construction process after the plant had begun construction on the site that would stop and we would have to redesign and move forward then. I think the certainty of the process is important so a plant can be built in a little over 3 years.

Senator LANDRIEU. Well, that brings me to my next question. Thank you very much. That was very helpful.

But, Dr. Travers, following up on that question, could you maybe state for the record some more specifics about what the NRC has done to facilitate the licensing and siting of nuclear power plants? Being optimistic in the sense that this energy bill will pass and we will get a green light, how much can you add to this statement of Mr. Bernhard's about how the process could go more smoothly, more safely, more efficiently, and less costly, which I think would be helpful to all concerned?

Dr. TRAVERS. Yes, thank you, Senator. I think a big step forward for the NRC was the promulgation of our relatively new—they have been existent for a while now—licensing regulations. Safety has always been job one, but in the past years, NRC has taken on a renewed commitment, I would say, to look at how we can be more effective in what we do, add predictability to the process.

Those new regulations have been successfully used, in part, to certify three designs now that can be referenced for any construction project for any organization that decides to actually build a plant. And all of the safety issues that were decided in connection with those certifications are essentially decided. They are put aside. They are not revisited unless some very high hurdle of safety concern is raised.

The other side of that equation is the environmental piece. As we have heard today, the nuclear industry is taking advantage of the possibility of obtaining early site permits. Those early site permits are another way of establishing predictability of the process. It allows you to argue that the site you would use ultimately, if you were to construct a nuclear power plant, is satisfactory. You obtain hearings in advance of spending any money to construct a plant. So you establish a predictability that you can bank that site for between 10 and 20 years, and ultimately the vision is that you could, when you wish to construct a plant, reference both the design certification and an early site permit to effectively allow you to come before the NRC once more but without revisiting all of the issues that were decided and resolved at those two points in time, and in a much more efficient process, license for construction and operation a new nuclear facility.

We have been actively attempting to test portions of our process. We have engaged the nuclear industry and the public with pieces of issues that have arisen and have been identified early so that we can reach resolution so that when the day comes, if it comes, we are prepared and we have the process in place that can be used most efficiently, again, with the focus principally on safety, to disposition that application.

Senator LANDRIEU. Can I follow up with one? That was very helpful, but let us talk about safety for a minute because that term had a certain definition before 9/11 and it has a different definition today. It is a very important issue that the people of our country are very focused on. I think it would be very helpful if you, Dr. Travers, would speak for just a moment about that. And if anybody else on the panel wants to take a shot at what would be the—you know, if asked at a reception or a party, are new nuclear power plants safe given post-9/11, would each of you take a minute and a half to go on the record with what you would say if asked? How would you answer that?

Dr. TRAVERS. I would be happy to. We think nuclear power plants were safe before 9/11. In fact, if you look, they were probably the most safe and secure, guarded commercial facilities in the United States.

After 9/11, though, legitimately some concerns were raised. The NRC took actions, working with the industry very cooperatively, very successfully I would have to say. We took a number of actions, and those actions have resulted in even a higher level of security and safety for nuclear power plants today. I could tick off a few specifics that relate to things such as the numbers of security personnel required at plants, the additional security posts, training for security members, time frame requirements for work hours, stand-off distances for vehicles, security checks, a whole host of issues including the background checks and tightened access requirements that have been put in place since 9/11.

We have been doing a lot. I would have to say the nuclear industry has been doing a lot, and today the view is that in the post-9/11 environment, nuclear power plants are even more secure and safe than they were before 9/11.

Senator LANDRIEU. Mr. McCullough or would anybody else like to add?

Mr. MCCULLOUGH. With your permission, Senator, I would like to ask Ike Zeringue, our president, to comment on that.

Senator LANDRIEU. Thank you.

Mr. ZERINGUE. We have essentially made it more difficult to access the general facility. We have hardened access to the plants. We have significantly increased our response time, and we have successfully performed a number of assault drills, special forces type assaults against the plant. There has been significant training there, increased background threats, and significant expenditure of resources to improve a variety of capabilities from training to armed response. We will spend on the order of an additional $30 million this year on improving security at our facilities.

Senator LANDRIEU. Thank you.

Mr. Fertel.

Mr. FERTEL. Senator Landrieu, I would echo what both Dr. Travers and Ike said, but let me just maybe take it industry-wide. We have increased the number of security officers from 5,000 to over 7,000. By the end of this year, we will have spent $1 billion across our industry. Ike mentioned $30 million at TVA. It will be

$1 billion across our industry, keeping in mind what Dr. Travers said, which was the plants were not only safe but had huge security requirements before 9/11. So we feel from a security standpoint, the plants are more secure today.

I think if there is an issue from a Senate standpoint or a national standpoint, it is how do you fit the nuclear plants, what we are doing into the critical infrastructure. We started off very secure. We are even more secure today, and when we are looked at, we are looked at almost in isolation.

What we worry about—in your State you have nuclear plants. You have a lot of chemical plants and fertilizer plants and they are all right next to each other. I have great confidence our nuclear plants are secure. I have less confidence, because I know less about it, that the other facilities in our critical infrastructure are nearly as secure. What we believe is necessary is to look across the spectrum of a critical infrastructure and make sure that both governmental and private resources are appropriately allocated.

Just one last point because you often hear in our industry that the industry is reluctant to do things, and I find it personally sometimes insulting because 1,000 people work at our plants, and if something was to go wrong at the plant, who at first is in jeopardy, but the people that work at the plant. Where do you think their families live? Around the plants. So no one wants to make sure that the plants are safer or more secure than the people that work there, and no one wants to make sure the assets are better protected than the owners and the management responsible for those.

So I think that the NRC has required us to do a lot of things, most of which may make sense, some of which we may not think make sense, but most does. We think you need to look broader to make sure that as a Nation we are really allocating resources going forward correctly to all our critical infrastructure.

Senator LANDRIEU. Well, I thank you all. I will just summarize by this, Mr. Chairman; as a supporter of this industry and getting it back up on its feet and revitalized and robust and moving forward, I think while the issues of safety that are raised and the concerns are most certainly legitimate, I would just cite to those here and listening the explosion of the tanker off the coast with a tank full of ethanol and the 98 percent of containers that come into this country that have absolutely no security, no check, no monitoring, and the kind of requirements that are necessary to keep those safe.

So I want, as an advocate, to just express for the record that the industry has made extraordinary steps to increase the safety and that you could almost argue that nuclear power may be among all the infrastructures, the safest part of that infrastructure. A, it can be identified. There are not 1,000 of them. They are specific sites. We know where they are. There are not going to be 1,000 nuclear power plants. The technology is there to protect it. I would argue that if we could get to that same level on chemicals, containers, railroads, tankers coming in and out of our ports, we would all be better off. So it is a red herring, and I hope that we can move forward and hopefully this energy bill give us that platform.

Senator ALEXANDER. Thank you, Senator Landrieu.

Let me direct a line of inquiry first to Mr. Fertel and then to Chairman McCullough.

I think the way you presented the environmental advantages to clean air of nuclear power is striking in your complete testimony. I want to make sure that I understand it right.

Mr. FERTEL. Sure.

Senator ALEXANDER. As I understand what you said in your complete testimony, your written testimony is that the nuclear power plants we have operating today in the United States, which produce about 20 percent of all the electricity we have, have avoided SO_2 emissions in an amount that exceeds the reductions imposed between the clean air amendments of 1990 and 2002. Did I get that right?

Mr. FERTEL. In just 1 year our——

Senator ALEXANDER. In just 1 year. 1 year's operation of those— excuse me. Go ahead.

Mr. FERTEL. In just 1 year, the operation of the nuclear plants in our country avoid about 3.4 million tons of SO_2 emissions. In the entire period from 1990 to 2001, the reduction in all of the SO_2 emissions from the rest of the fossil generation source was 5 million tons. So in 1 year, we do 3.4. It took 11 years to reduce 5 million tons from all the fossil units. Now, that is not saying they are doing bad. It is just saying that nuclear has a very, very significant impact on achieving clean air goals in our country.

Senator ALEXANDER. And then you go NO_X, nitrogen. We are talking basically about soot and smog of the kind we increasingly see in our part of the world. You say that the NO_X emissions avoided by the power plants are equivalent to eliminating NO_X emissions from 6 out of 10 passenger cars.

Mr. FERTEL. That is correct, sir.

Senator ALEXANDER. And you say that the carbon emissions avoided by the nuclear power plants we have operating today are equivalent to eliminating the carbon emissions from 9 out of 10 passenger cars.

Mr. FERTEL. That is true too, and in our voluntary carbon reduction program, nuclear energy makes up about 45 percent of the total voluntary reductions for our Nation from every industry, not just the electric power industry.

Senator ALEXANDER. I wonder if you have considered, as we talk about various clean air proposals in the Congress, say, taking the President's Clear Skies proposal or Senator Carper's proposal, which I am a cosponsor of, which is a little tougher on NO_X and SO_2 and adds carbon, and comparing that to nuclear power. I think it is very helpful and it helps make the case of why it is a useful alternative.

Mr. FERTEL. We could look at making a comparison. Unfortunately, nuclear on clean air is kind of like electricity for us as Americans. When we did some focus groups years back on where does electricity come from, what we got back from the focus group were the switches and the outlets. So if I wanted more electricity, I would put in more switches and more outlets. Now, when you had more discussion, they realized power plants were somehow involved, but electricity is taken for granted, except when we miss it. So is clean air.

Senator ALEXANDER. Well, when people have to start driving at 55 miles an hour and stop cooking in their back yards and going into some place and standing in line to get a car emissions sticker, then they begin to pay a little bit more attention, which is what is about to happen all across Tennessee.

Mr. FERTEL. Yes, sir.

Senator ALEXANDER. If I may switch over to Mr. McCullough. TVA's coal fleet, which is most of the power—what percent of TVA's power production is coal?

Mr. MCCULLOUGH. About 56 percent, Mr. Chairman.

Senator ALEXANDER. And my information is that the coal fleet is 40 to 45 years old. It is the oldest coal fleet in the Nation. Would that be right?

Mr. MCCULLOUGH. I would have to do some research, but 47 is the average age. You are right.

Senator ALEXANDER. How much does it cost to put a scrubber on a coal-fired power plant smokestack? I know I hear TVA is spending $1 million a day on air pollution. What is the cost of that?

Mr. MCCULLOUGH. Approximately $250 million per scrubber.

Senator ALEXANDER. Per scrubber.

Mr. MCCULLOUGH. Yes, sir.

Senator ALEXANDER. Yet, I read the other day that TVA has just signed a 20-year contract to buy $3 billion of high sulfur coal. Why does it make sense to buy high sulfur coal and then pay $250 million per scrubber to scrub out the sulfur?

Mr. MCCULLOUGH. It is a matter of the capital cost to replace the generation. We can install a scrubber and we use a detailed model to determine which is the lowest cost, environmentally compliant baseload generation source.

Senator ALEXANDER. But what I am getting at is why not buy low sulfur coal.

Mr. MCCULLOUGH. We do buy low sulfur coal to the extent that we can.

Senator ALEXANDER. Why would you buy $3 billion of high sulfur coal over a 20-year period of time when 80 percent of the State is about to be in violation of the Federal Clean Air Act?

Mr. MCCULLOUGH. May I ask Ike Zeringue to respond to that?

Senator ALEXANDER. Yes.

Mr. ZERINGUE. We cannot meet the requirements of the Clean Air Act by simply replacing high sulfur coal with low sulfur coal at our plants. As a result, we are putting scrubbers on these facilities to remove the sulfur from those plants, and we will have removed approximately 75 to 85 percent of the sulfur emissions by 2010.

Senator ALEXANDER. But would it not cost more to remove sulfur from high sulfur coal than from low sulfur coal?

Mr. ZERINGUE. No, sir.

Senator ALEXANDER. It does not?

Mr. ZERINGUE. No, sir.

Senator LANDRIEU. Why is that?

Senator ALEXANDER. I will take your word for it, but that sounds odd to me.

Senator Landrieu, may I ask one more question?

Senator LANDRIEU. Oh, sure.

Senator ALEXANDER. And then I will go to you for whatever time you would like.

What I would really like to get at is if there is such an advantage to nuclear power in terms of clean air—and in the Knoxville area especially in the Great Smoky Mountains, which is a class 1 protected area, we have a difficult problem and we have it in terms of ozone and we are going to have it next year in terms of particulate matter. And I salute you for this decision you have made to reopen Browns Ferry.

Why would it not make sense to move more rapidly if the Browns Ferry opening stays on cost and on schedule, as it is today and as we hope it will be? Why would it not make sense to move more rapidly to reopen Watts Bar and perhaps Bellefonte and close the most offensive of the coal-fired power plants which are producing so much nitrogen and sulfur emissions, as well as carbon?

Mr. McCULLOUGH. Yes, sir. We evaluate the fuel costs and we evaluate also TVA's responsibility to cleaner air. We evaluate a diverse portfolio that we have to maintain against the capital cost that would include either a scrubber or the capital cost to bring on line Watts Bar 2 or to bring on line a unit at Bellefonte. My answer, Senator, is that it is more cost effective and environmentally responsible for us to invest in scrubbers, and we do buy as much low sulfur coal as we can. There is an economic model that determines really the optimal point.

But I think in the future you will see the expansion of safe, reliable nuclear, but there is economics involved there and also a fuel supply cost involved there. We will be happy to sit down with you and go into more detail.

Senator ALEXANDER. I would look forward to that. I know that you make long-term plans. I do not intend to try to be the Chairman of the Tennessee Valley Authority or a manager, but it would seem to me that along that theory, unless a boiler blows up, you are going to be operating these coal-fired power plants forever. What I am wondering is whether it might not be wiser to accelerate nuclear power development and begin to close the most offensive of those plants. But we will talk about that at a later time.

Thank you, Senator Landrieu, for letting me pursue that.

Senator LANDRIEU. Thank you, Mr. Chairman. I most certainly did not mean to interrupt. This is your area, TVA.

But I am just curious. Do we get all of our coal locally, the coal that we were just speaking about? Or does some of it come from other places in the world?

Mr. McCULLOUGH. The coal that TVA burns is domestically mined.

Senator LANDRIEU. All of it?

Mr. McCULLOUGH. Yes, ma'am.

Senator LANDRIEU. Thank you. That is all I wanted to know.

Senator ALEXANDER. Do you have any other questions?

Senator LANDRIEU. No.

Senator ALEXANDER. We have a vote at 4 o'clock and it is about time to bring this to a conclusion anyway.

This has been very useful testimony. Your written statements are extremely helpful.

The purpose of this hearing has been to talk about the future of the nuclear industry, to highlight the impediments and the opportunities there. It, of course, had an opportunity to highlight what the Tennessee Valley Authority is doing with the Browns Ferry plant, which the whole country is watching, which the Congress will watch very closely.

I am glad to see that the Nuclear Regulatory Commission over the last several years has been able, while we are moving in a safe and efficient way, to create a clearer path from beginning to establishment of a new plant, one of which we have not had in 30 years.

We heard from Mr. Asselstine what his six or seven criteria are for private investment coming back. We have heard about the environmental advantages, and we have heard that France and Japan and India and Russia are all taking a technology we invented and doing much more with it than we are able to.

So perhaps we have reached a point where we will have steady progress toward the revival of the nuclear power plant as a way of help keeping jobs in America, not moving overseas, help cleaning our air and providing reliable, efficient energy at a reasonable cost.

Thank you very much for coming.

[Whereupon, at 3:58 p.m., the hearing was adjourned.]

APPENDIX

RESPONSES TO ADDITIONAL QUESTIONS

March 25, 2004

Mr. GLENN L. MCCULLOUGH, JR.,
Chairman, Board of Directors, Tennessee Valley Authority, Knoxville, TN.

DEAR MR. MCCULLOUGH: I would like to take this opportunity to thank you for appearing before the Senate Committee on Energy and Natural Resources hearing on March 4, 2004 regarding New Nuclear Power Generation in the United States.

Enclosed herewith please find a list of questions which have been submitted for the record. If possible, I would like to have your response to these questions by April 8, 2004.

Thank you in advance for your prompt consideration.

Sincerely,

PETE V. DOMENICI,
Chairman.

RESPONSES TO QUESTIONS FROM SENATOR BUNNING

Question. TVA is over $26 billion in debt. This is very close to the $30 billion debt cap. In 1997, TVA promised, in exchange for a rate increase, to lower its debt level by half from $26 billion to $13 billion. Seven years later, TVA has barely reduced its debt. TVA has proposed, however, restarting its Unit 1 Browns Ferry facility at a cost of approximately $1.8 billion. Why does TVA believe that restarting Unit 1 is a prudent business policy? How does TVA believe restarting Unit 1 will better serve the TVA customers?

Answer. TVA will need the base load generation that Browns Ferry Unit 1 will provide when it is restarted in May of 2007. Demand for electricity in the Valley is projected to grow at 1.8% over the next five years. The region has grown at a faster pace than the national average. With the help of independent analysts, we spent almost two years investigating the prudence of restarting Unit 1 at Browns Ferry and determined that it was the most cost effective way of achieving our new base load needs. The cost of the restart was evaluated against additional purchases from IPP's, coal gasification, etc. In fact, our current resource-planning analysis shows that this nuclear unit will help us meet our growing energy needs at a very competitive cost by reducing our delivered cost of power by about .09 cents per kilowatt-hour in its first year of operation, and the unit is expected to have paid for itself seven years after start-up.

Although there is currently an excess capacity of generation caused by the construction of gas plants in or near the TVA region, this does not change the economic advantage of restarting Browns Ferry nuclear unit 1. At gas prices of $4 $6mmbtu, the total cost of power produced by the most efficient of these new plants is $35 $50 megawatt hour. In contrast, the total cost of Browns Ferry 1 will be near $26 megawatt hour.

Question. Under current law, TVA may only serve power to its own region and may not sell electricity outside its "fence". It is my understanding that TVA already has a 40% surplus supply of power within the "fence". If TVA cannot sell outside its fence, why does TVA believe it is necessary to increase their debt by restarting Unit 1 Browns Ferry to increase its supply of power?

Answer. TVA is taking advantage of the opportunity to purchase power from IPP's in the region when it represents the best business decisions for consumers. We have contracts with Tractebel and Calpme.

When TVA was considering the restart of Browns Ferry Unit 1, it carefully considered alternatives for purchasing necessary baseload capacity and determined that restarting Browns Ferry Unit 1 added to the strength of our generation mix and lower TVA cost of power.

For the summer of 2004, TVA's own reserve margin will be 13%—a highly appropriate reserve margin under industry standards to ensure reliable supply of power to the TVA region. When Browns Ferry Unit 1 is returned to service in the summer of 2007, it is anticipated that TVA's reserve margin will be 13.2%, demonstrating that Browns Ferry Unit 1 is devoted to meeting the needs of TVA's power customers inside the "fence".

Question. For a fraction of the cost of restarting Unit 1 Browns Ferry, TVA could upgrade its electricity grid and interconnect it with some of the many new clean burning gas plants that have been constructed in the last 5 years and receive twice as much power as can be produced by a 40 year old nuclear power plant design. Has TVA done a cost benefit analysis of importing power from other sources to meet their electricity needs compared to restarting a 40 year old nuclear plant?

Answer. Although there is currently an excess capacity of generation caused by the construction of gas plants in or near the TVA region, this does not change the economic advantage of restarting Browns Ferry nuclear unit 1. At gas prices of $4 $6mmbtu, the total cost of power produced by the most efficient of these new plants is $35 $50 megawatt hour. In contrast, the total cost of Browns Ferry 1 will be near $26 megawatt hour.

In making this decision, TVA studied several potential options, including purchase power prices, combined-cycle gas turbines, coal gasification, startup of one of the Bellefonte nuclear units, and the recovery of Browns Ferry nuclear unit 1. Each potential option was studied in terms of fuel price stability; long-term cost to produce power; environmental impact; potential impact to TVA's long-term ability to reduce debt; capital cost; and estimated capacity factor for meeting baseload needs. Numerous sensitivity studies were conducted on each option, utilizing TVA and industry forecast data to compare each to forecasted future electricity, gas, and uranium prices. Results of these studies indicated that recovery of Browns Ferry Unit 1 would be very favorable to meeting TVA's baseload requirements.

Unit 1 is a viable asset that will effectively and efficiently provide clean, affordable and reliable power to meet the future power demands of people in the Tennessee Valley. We're refurbishing Unit 1 in the same manner that we accomplished for Units 2 and 3, and they are both running well.

Browns Ferry Unit 1 has been maintained in its present configuration using long-term lay-up programs monitored and inspected by the Nuclear Regulatory Commission. A thorough analysis of the scope of work we needed to do to ensure safe, reliable operation was performed and we are implementing that scope of work. These results also indicated that it would be able to produce the needed energy at very competitive rates as compared to the other available options, while maximizing TVA's investment in an existing asset.

Question. TVA continues to tell Congress that it will reduce its debt. It has increased my Kentucky constituents' power rates in an effort to reduce its debt. However, the debt still remains large at $26 billion and my constituents who use TVA power have some of the highest electricity rates in Kentucky. Outside of the nearly $1.2 billion that the federal government forgave of TVA's debt, how close are you to meeting your goal of cutting your debt in half from $26 billion to $13 billion?

Answer. No portion of TVA debt's has been forgiven by the federal government, but $3.2 billion of long-term debt held by the Federal Financing Bank was paid off in full and refinanced at lower interest rates in FY 1999, thus saving substantial interest costs. Further, the U.S. Treasury receives payment each year on the government's original investment of $1.4 billion dollars plus interest. To date, approximately $3.5 billion has been repaid.

The assumptions used to establish the debt reduction target in the earlier 1997 plan did not take into account any Clean Air expenditures beyond those known and budgeted at the current time or the debt impacts of for building new base load and peaking generation (power needs were accounted for; it was assumed that it would be purchased, as it is in the current plan beyond building Browns Ferry). Nonetheless, TVA reduced its debt by approximately $2.9 billion from FY 1997 through FY 2003. As you know, TVA completed a transaction MLGW to prepay $1.5 billion of its future power purchases; TVA used those proceeds to reduce statutory debt. This transaction is not included above as this was a FY 2004 transaction. Our recently released Strategic Plan identifies a debt reduction target of $3 to $5 billion over the next 10 to 12 years.

○

APEX
TP_CF
9780160717406